S0-EIB-968

Laboratory Minicomputing

Notes and Reports in Computer Science and Applied Mathematics

1. John R. Bourne. Laboratory Minicomputing
2. Carl Tropper. Local Computer Network Technologies

Laboratory Minicomputing

JOHN R. BOURNE
Department of Electrical and Biomedical Engineering
Vanderbilt University
Nashville, Tennessee

1981

ACADEMIC PRESS
A Subsidiary of Harcourt Brace Jovanovich, Publishers

New York London Toronto Sydney San Francisco

ACADEMIC PRESS, INC.
111 Fifth Avenue, New York, New York 10003

United Kingdom Edition published by
ACADEMIC PRESS, INC. (LONDON) LTD.
24/28 Oval Road, London NW1 7DX

Library of Congress Cataloging in Publication Data

Bourne, John R.
Laboratory minicomputing.

Includes bibliographical references and index.
1. Minicomputers--Programming. 2. PDP-11
(Computer)--Programming. I. Title.
QA76.6.B684 001.64'2'0245 80-70668
ISBN 0-12-119080-3 AACR2

PRINTED IN THE UNITED STATES OF AMERICA

81 82 83 84 9 8 7 6 5 4 3 2 1

Contents

6. I/O Fundamentals

7. Laboratory I/O—A/D, D/A, Clocks

8. Interrupts and Real-Time Programming

9. Example Programs

Preface

Minicomputers have become almost indispensible laboratory tools in engineering, the medical and physical sciences, and other disciplines, including various areas in the liberal arts. While general information about computer programming is abundantly available, most users of laboratory minicomputers will find it difficult to secure information specifically tailored to their needs. Common laboratory tasks such as acquiring analog data, displaying data, and controlling real-time activities are infrequently explained in text and tutorial information. The purpose of this book is to provide such information. Background material is provided to allow the reader with little or no knowledge of minicomputers to learn to successfully use minicomputers for a variety of laboratory data acquisition and analysis purposes.

The hardware systems described are the PDP-11 minicomputer and LSI-11 microcomputer, both manufactured by Digital Equipment Corporation. The UNIX* operating system and the 'C' programming language are used as primary software tools in examples throughout the text. The UNIX system and 'C' were chosen because of their simplicity, ease of use, and widespread adoption in the university community. These software tools are used throughout the Bell System and also by other industries.

* UNIX and MINIUNIX are trademarks of Bell Laboratories and are available under license from Western Electric, Greensboro, North Carolina.

The reader who works through this book should be able to write programs for PDP-11s, using either 'C' or assembly language in most conceivable laboratory settings. The positive experiences of some 200 students who have studied this material over the past five years testify to its utility.

The three appendices in this text contain

(1) a discussion of methods for bringing up the UNIX system on PDP-11s,

(2) a description of how to modify the MINIUNIX system to run on PDP-11/03s, and

(3) a list of program listings available as a supplement to this text. This supplement is entitled "Program Listing Supplement for Laboratory Minicomputing." The programs include the code for intercomputer communications, various device handlers, and other useful programs. The device handler code can be distributed only to holders of a UNIX license. The source code for the above programs is available on a 9-track tape from the author for a nominal handling and materials charge.

The author would like to express his appreciation to Mr. Gary Woyce, Mr. Danny Johnston, Mr. Eric Blossom, and Mr. Jim Cannon for their help in reading the manuscript and testing the examples and problems. Thanks also go to Dr. Douglas Giese and Mr. Bruce Giese for work on example programs while they were students at Vanderbilt University. Partial support for the development of the spectral analysis programs in Chapter 9 came from the National Institutes of Health and is hereby acknowledged. The author would like to thank Ms. Jacqueline Boswell for her typing of the manuscript, his wife for her diligent proofreading of the original manuscript, and Dr. A. J. Brodersen for his continued encouragement and administrative support of the project.

1

Introduction

1.1 OVERVIEW

This is a book about laboratory minicomputing, specifically concentrating on PDP-11s running the UNIX* operating system. The PDP-11 series of minicomputers is manufactured by Digital Equipment Corporation (DEC) in Maynard, Massachusetts, and the UNIX system is an operating system developed by the Bell Telephone Laboratories for use on a number of different computers including the PDP-11. Although many basic concepts are described in this text that are applicable to different types of computers, the reader who is able to test out the examples given and perform the exercises suggested on a PDP-11 will be able to progress significantly faster than the reader without a computer. The PDP-11 family was chosen as an example machine because the 11 family is widely used in university, industrial, and governmental laboratories. Moreover, besides being a powerful and general system amenable for use in various laboratory settings, the PDP-11 is also available in the hobby market (Heathkit, for example). The UNIX operating system and the language 'C' were chosen for use in this text because of their wide use in university settings and the simplicity and ease of understanding these software tools. Further, the 'C' language (a top-down structured language) is not unlike Pascal and is currently in use in many scientific and commercial markets.

The main purpose of this book is to enable the reader to write programs to use common peripherals found in most laboratories [e.g., analog-to-

* UNIX is a trademark of Bell Laboratories.

digital converters (A/D), digital-to-analog converters (D/A), real-time clocks, parallel digital I/Os, etc.]. The experience gained with the specific systems described in this book can be generalized to other machines and operating systems. Real-time programming using various peripherals will be described in detail in the latter chapters of the text with illustrative examples.

1.2 HISTORICAL PERSPECTIVE

Although small computers were used in laboratory settings as early as the late 1950s, widespread laboratory use did not occur until the late 1960s and 1970s when vastly reduced prices permitted most educational institutions, industrial concerns, and governmental laboratories to install minicomputers in large numbers. One of the first small computers, marketed in 1960, was the PDP-1 by Digital Equipment Corporation. The cost, at that time, was only 10% of larger competitive machines. The PDP-5, introduced in 1963, later led to the PDP-8 series. This series survives into the 1980s with new machines being sold on a regular basis. The PDP-8 has enjoyed surprising longevity, primarily because of a large continuing software effort. Programs already written for PDP-8s encouraged many computer users to purchase PDP-8s even though more sophisticated minicomputers such as the PDP-11 became available after 1970. The PDP-8 is a 12-bit word computer. Initially offered with 4096 words of memory, most PDP-8s now use larger memories up to a maximum of 32,000 words. As costs for memories decreased, users typically added additional memory to their systems. PDP-8s used in many laboratories were initially sold with the label LINC-8 and later, the Lab-8. The LINC-8 was a two-processor machine, containing a LINC and PDP-8 processor (Bell and McNamara, 1978). Typically, these names indicated that a system consisting of a computer, tapes, and laboratory analog-to-digital and digital-to-analog units were packaged together. Early costs for relatively small systems were in the range 30–50 thousand dollars. Now, comparable systems are available for well under $10,000.

As costs for computer systems have dropped over the last two decades, user's appetites for more and more powerful capabilities have risen, thereby keeping expenditures for computer-related equipment high. Even in the hobby market, an expenditure of one to three thousand dollars will provide the home enthusiast with a system capable of carrying out productive work. Early computer systems often consisted of only a central processing unit (CPU) and one or more input/output (I/O) devices, typically a terminal and, for laboratory applications, an A/D and D/A. Pro-

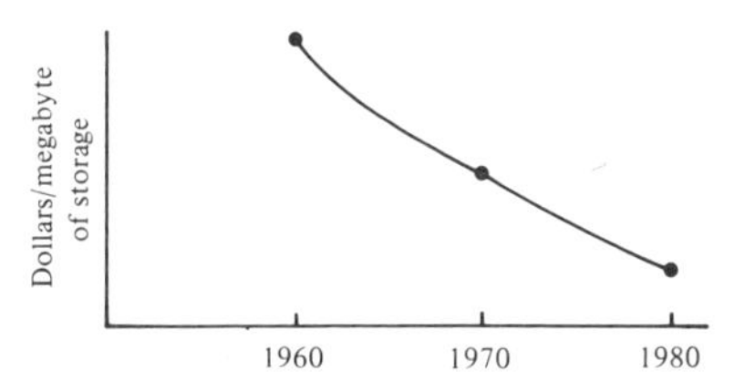

Fig. 1.1. Cost/million 8-bit (1-megabyte) characters stored.

grams and data were frequently stored on paper tape and read into the computer via a terminal 10 character/sec paper tape reader or high-speed paper tape reader. Less frequently, programs were loaded by hand into the computer via a set of switches (the switch register). Cumbersome methods were used in early small computer systems. For example, to produce a running Fortran program on a typical small system without mass storage, the user was required to generate a paper tape of the source code [i.e., the text of the Fortran (or other language) program], feed the tape reader for two or three passes with the original and intermediate tapes before a running program was produced. This procedure would typically require hours for all but trivial programs.

The advent of mass storage devices (e.g., disks, magtapes) for minicomputers made the use of paper tape obsolete. A popular medium in the late 1960s and early 1970s for DEC computers was the DECtape unit, a small magnetic tape that could hold up to 262,000 characters of information. Digital tape cassettes for recording achieved high popularity because of low cost. Small but high-speed disk drives with inflexible (hard) recording platters (similar in size to phonograph records) were initially very expensive. Figure 1.1 indicates conceptually the relative cost of disk storage versus year in dollars/megabyte (1 megabyte = 1 million characters) of storage. Until after the mid-1970s hard disk storage capacities for minicomputers were generally less than 2 megabytes/disk cartridge. In the late 1970s larger disks became available at costs not dissimilar to the costs of previously available drives. For example, Fig. 1.2 shows conceptually

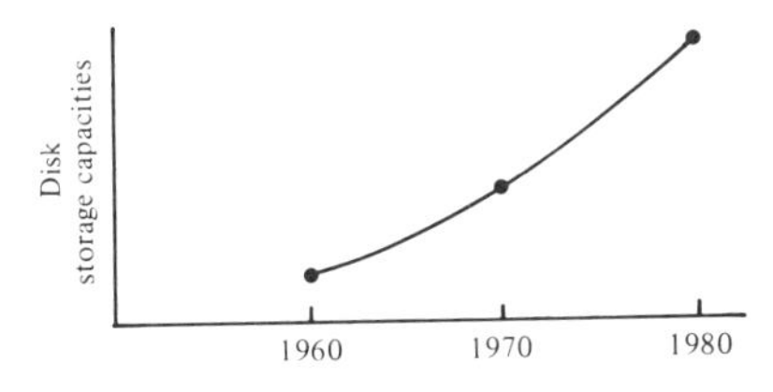

Fig. 1.2. Disk capacity versus time for constant cost.

how constant dollars could purchase larger and larger amounts of disk storage in later years.

In addition to "hard" disks, "floppy" or small flexible disks became widely used in the mid-1970s and are currently perhaps the most employed inexpensive storage medium. A floppy disk may cost as little as a few dollars and contain up to a megabyte of storage. In contrast, hard disk costs range from somewhat less than $100 to several hundred dollars for storage capacities ranging from 2 to 5 megabytes up to several hundred megabytes. The larger storage capacities are achieved by using cartridges that contain several vertically stacked platters.

The time required to access data from any location on a disk platter is much faster for a hard disk than for a floppy disk due to higher rotational speed of the hard disk and to faster movement of the head(s) that read the information from the disk surface. Consequently, the overall performance of minicomputers using floppy disks is much worse than computers running with hard disks. However, floppy disk units are much cheaper than hard disks. In making a decision to buy either floppy or hard disk units, the user must carefully examine the tradeoffs between cost and speed.

1.3 USES OF MINICOMPUTERS

It is no understatement to say that minicomputers (perhaps in smaller versions using microprocessors) will pervade the entire fabric of society in the not too distant future. We are all familiar with popularized discussions of uses of computers in business, the stock market, automobiles, medicine, hobby activities, banking, etc., but are perhaps less familiar with how small computers are actually connected to situations involving the measurement and control of devices in the world outside the computer. This volume will attempt to investigate this interface area in a specific context of one currently popular minicomputer—the PDP-11. No attempt will be made to generalize the information presented to a host of application areas since they are legion. Instead, given the concepts presented, the reader should be able to apply them to his own environment. Nevertheless, an introductory word about use of minicomputers in laboratory-type situations is appropriate.

In engineering, minicomputers are involved in a host of activities ranging from automated circuit design, machine and process control, data acquisition, analysis, display, and management. In medicine and science, minicomputers are routinely used in laboratories for control of experiments, data acquisition, and display, for patient monitoring, and in the operating room. The next sections will describe first the conceptual basis and component elements of minicomputer systems and then give several examples of how such systems are used in a number of selected examples.

1.4 COMPONENT ELEMENTS OF A GENERAL LABORATORY MINICOMPUTER SYSTEM AND EXAMPLES

Figure 1.3 shows a block diagram of typical elements in a basic laboratory minicomputer system. The CPU (central processor unit), containing control facilities and logic/arithmetic capabilities, is typically attached to a variety of peripheral devices. These devices, called "peripherals" because they are not central, include all hardware added to the basic computer, such as disk drives, analog-to-digital converters, etc. The figure shows some popular peripherals and how they interface to (read plug into) the CPU. The amorphous mass at the bottom left of the figure is labeled "monitor," indicating the presence of either a human observer who watches an oscilloscope for example, or a printer or other type of logging device. The main feature of this system is the interface to the experiment or process (i.e., the real world). Common interface needs require analog data acquisition [analog data via the analog-to-digtal converter (A/D)], digital data sensing (e.g., a switch) via the parallel digital I/O (DI/O), and returning data to the external world via the digital-to-analog converter (D/A) and DI/O (e.g., opening or closing a relay). Data acquired can be stored on a mass storage medium (e.g., disk, tape), sent somewhere else

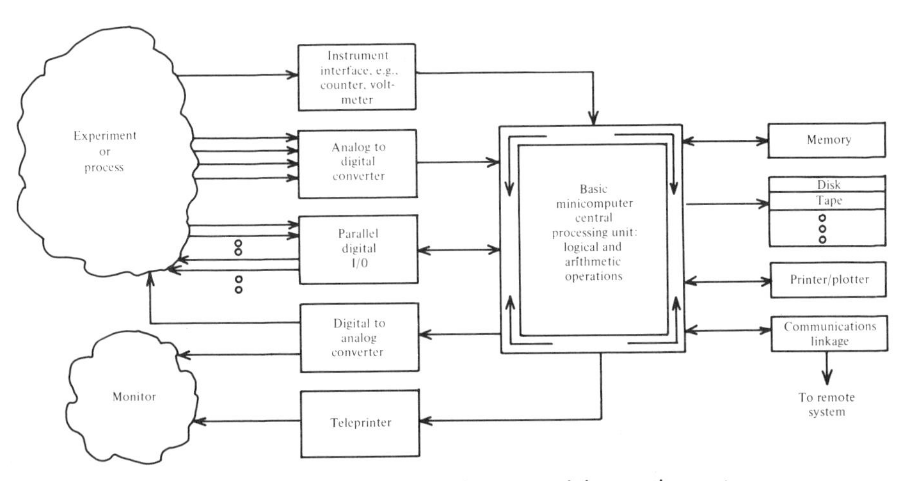

Fig. 1.3. Typical elements in a laboratory minicomputing system.

via telephone or cable, printed out or observed on a display device. Memory is used for both storing data and programs.

Many examples of the use of laboratory minicomputer systems can be given. Whether such systems are used in biological, electrical, mechanical, or other applications is basically immaterial in the context of this book. The concepts employed in most applications are similar if not almost identical. Consider the common procedure of acquiring data via the A/D, Fourier transforming that data, and displaying the power spectra on an oscilloscope or plotting the result on an x–y plotter. The ability to conduct this type of experiment is desired in many laboratory settings in which knowing the frequency content of a signal is the required end product. For example, to name a few:

1. *Speech analysis.* The study of frequencies in a speech signal may provide audiologists with a tool for analyzing abnormalities in speech.
2. *Fracture analysis.* For example, attaching strain gauges to airplane wings and observing the periodic frequency components during flight may allow prediction of failures.
3. *Determining roundness of automobile tires.* This is an easy example to understand. Some years ago a major tire manufacturer wanted to determine the roundness of his tires. A test situation was set up as shown in Fig. 1.4. As the tire rotated, an oscillation in the amount of light falling into the photocell would be observed if the tire was not perfectly round. An example of the output generated is shown in Fig. 1.5. In the case of a defective tire, the periodicity of the waveform could be related to the

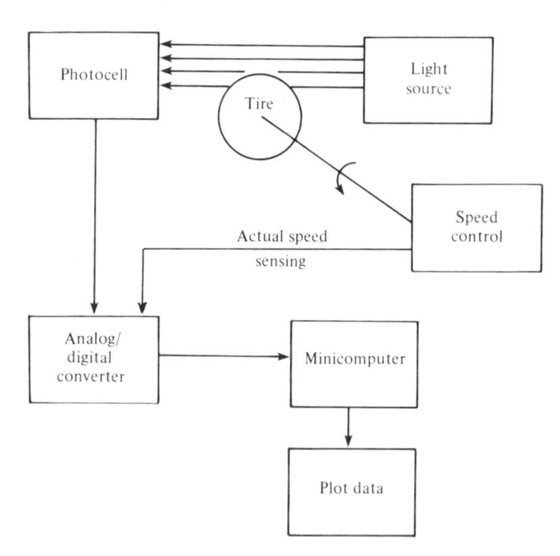

Fig. 1.4. A method for testing the roundness of automobile tires.

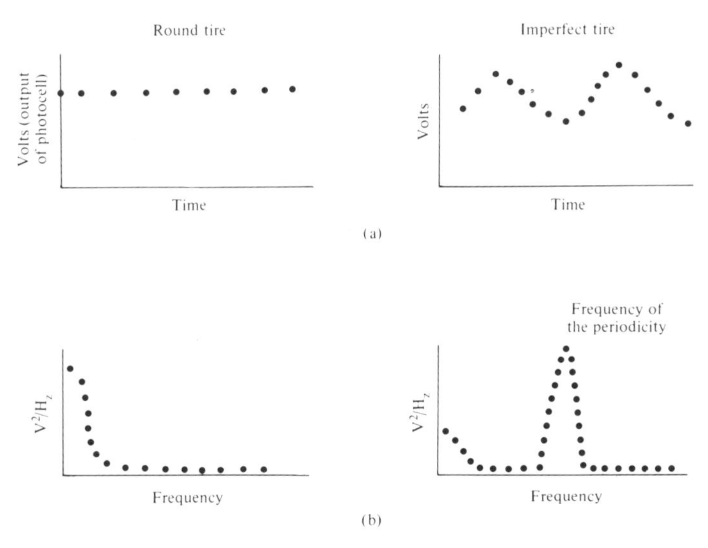

Fig. 1.5. Data plots produced by tire testing system: (a) digitized data, (b) spectral data.

defect if the speed of tire rotation was known. Consequently, the persons responsible for testing the tires fed the digitized waveform into a Fourier transform/spectral analysis program in a minicomputer to determine the frequency of the periodicity. If the periodicity was equal to the speed of rotation, the tire had one imperfection. If the waveform, however, had several frequencies present, considerable evidence was provided that more than one imperfection was present.

4. *Spectral analysis of electroencephalograms (EEGs).* This is probably the most common method in use in the world for analyzing EEGs or brain waves. Most laboratories employ spectral analysis to conduct research on brain wave characteristics. The technique is also in clinical use in many hospitals.

Many other common procedures are routinely carried out with laboratory minicomputers including:

1. *Signal averaging.* These procedures are used to extract periodic signals from noise.

2. *Histogram analysis.* Creation of histograms of the time of occurrence of events after a stimulus versus the number of events is a common analysis procedure used in psychology and neurophysiology for studying the nervous system and behavior.

3. *Experimental and industrial control.* Numerous control examples can be given ranging from automated control of automobile assembly to

control of behavioral testing procedures. While industrial automation procedures may not strictly fall under the heading of "laboratory" procedures, the methods used for implementing industrial systems are essentially the same as the procedures used in implementing laboratory systems.

Experimental control systems such as monitoring of sensors in solar heating systems or controlling banks of switches that are used for presenting various stimuli to subjects are easy to implement using small laboratory minicomputers. One example from the theater industry is the control of lights in a programmed sequence in a theater performance. Similarly, the music industry has begun to use minicomputers and microprocessors for such activities as instrument tuning and producing artificial/special effects music.

4. Minicomputers are routinely employed in other industrial applications such as data acquisition from oil and gas wells, testing of propellers on boats, reading bank checks, analyzing blood with automated analyzers, etc. The areas in which one can acquire analog data, analyze the data, and feed information back for either control or observation are so broad that it is impossible to provide a comprehensive list of applications. One suspects that the use of mini/microcomputers attached to the external environment will expand greatly in industrial, academic, and governmental environments in the years to come.

Computers are more useful to computer users when it is easy to communicate with the computer. The next section describes some basic concepts in *operating systems,* which provide a software interface between the computer hardware and the user.

1.5 ELEMENTARY OPERATING SYSTEM CONCEPTS

The software written to interface the computer user to the computer is called an operating system (i.e., it allows the user to 'operate' the computer). Early minicomputers did not have operating systems, and programs were loaded directly into memory via switches or paper tape as shown below:

Step 1 (a) Generate a paper tape with ASCII code (characters) or
(b) toggle the program machine code directly into computer with switches on the front panel; if toggled into memory, go to step 3.

Step 2 Assemble the program entered on paper tape.

Step 3 Run program.

Such cumbersome procedures provided the impetus for the development of minicomputer operating systems. Minicomputer manufacturers initially developed operating systems for single users and later for multiple users. For the PDP-8, the single-user system developed was OS-8 (*o*perating *s*ystem-8) and for the PDP-11, RT-11 (*r*eal *t*ime-11). OS-8 allowed only one user to use the PDP-8 system at a time. RT-11 was also similarly restricted but could also run in a foreground/background mode in which two programs contemporaneously ran. That is, the foreground program would run whenever it needed to (for example, whenever an A/D sample was taken) and the background program would run the rest of the time.

Multitasking systems were developed (e.g., RSX-11), which allowed more than two programs to run contemporaneously, each with a different priority. Time-sharing systems were also developed (e.g., RSTS) that

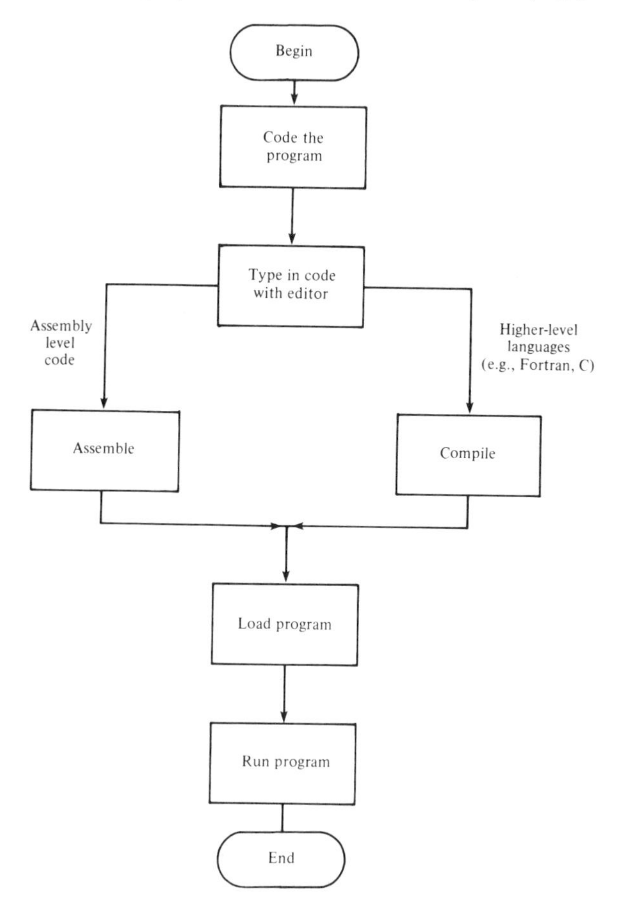

Fig. 1.6. Procedures for producing a running program.

allowed multiple users a slice of the available time on a "round-robin" basis. While each minicomputer manufacturer developed its own software system, Bell Laboratories developed a time-sharing operating system called UNIX for its inhouse use. The UNIX system is used in examples throughout this book and has become the major operating system in our department at Vanderbilt University. The UNIX system is available under license from Western Electric at no charge to educational institutions. Its simplicity, ease of use, and modifiability make it very useful for student instruction. Further, it provides an excellent medium for producing programs for both large PDP-11s, small PDP-11s, microprocessors, and other computer systems running UNIX systems (e.g., IBM, Interdata, Honeywell).

In most operating systems, common steps are taken to produce running programs. Figure 1.6 shows the usual sequence of operations that a user must follow. First, the program is coded (written down or typed into the computer directly). Suppose, without knowing anything about the UNIX system or the 'C' language, that we follow the steps shown in Fig. 1.6 for running the following program

```
main() {printf ("Hi There!");}
```

First, the program is typed in (underlined characters typed by computer):

```
% make hi_there.c
* imain () {printf ("Hi There!");}
$ ex$$ ($ = escape key)
%
```

Next, compile and load,

```
cc hi_there.c
```

Move output to a permanent name

```
% mv a.out hi_there
```

and run

```
% hi_there
Hi There!
```

In brief, the program is typed in using an editor (accessed here by the command "make"). The program is converted (compiled) into a series of machine instructions by the compiler "cc." The machine instructions are left in a file called a.out and renamed "hi_there" before running the program. This example is given primarily for flavor and the reader is referred to Chapter 5 for details on the procedure.

In Fig. 1.6, the left branch in the flow chart marked "assembly level code" is the route taken for conversion of simple symbolic instructions, closely related to the actual machine bit patterns, to a set of code that will actually run on the computer. For example, there may be various basic instructions in a computer (e.g., add, subtract, etc.) that each correspond to a set of bits arranged in a computer word. The purpose of the assembler is to convert these basic instructions into bit patterns that the computer can recognize. The basic instructions represented by simple terms such as "add," "subtract" are called mnemonic instructions and code in this form is referred to as symbolic code.

Assembly level code is typed into the computer just like Fortran and other higher level languages. When assembled it produces a stream of bits that can be loaded into computer memory. Some compilers produce assembly level code as an intermediate step to machine code. In these cases, the diagram in Fig. 1.6 would show compilation and then assembly prior to loading. An example for assembly level code is given below in a procedure that is similar to the compilation method shown for 'C' code:

```
% make add          [create a file called "add"]
I add $1, a         [a = a+1, the value one is added
                    to a]
  a: 0              [reserve memory location for a]
  $ex$$
% as add            [assemble and load into a file
                    called "a.out"]
% mv a.out add      [move a.out to file called "add"]
% add               [program runs]
%
```

The program adds the value 1 to a variable named "a" where "a : 0" reserves a memory location for "a" and defines "a" initially to have a value of zero. The statement "as add" will convert the symbolic statements to machine code and leave them in a file named "a.out" which is then renamed "add." When "add" is typed, the program runs and "1" will be added to "a" in the computer's memory.

Most operating systems provide similar sets of commands to the user. Table 1.1 shows some of the common facilities provided to the user by the UNIX system. Files containing characters may be produced by any of the various available editors. Files are composed of text (e.g., programs

TABLE 1.1

SOME COMMON FACILITIES AND PROGRAMS PROVIDED TO THE COMPUTER USER BY A TYPICAL UNIX SYSTEM

Description	UNIX Command
1. List directory	% ls
2. Make and edit files	% ed (UNIX editor)
	% teco (text editor and corrector)
	% edit (RT-11 editor)
3. Compile files	% cc 'C' files
	% fort RT-11 Fortran (Princeton version)
	% fc UNIX Fortran
4. Assemble files	% as
	% macro
5. Produce output	% cat concatenate and type
	% print

typed in, books, etc.), programs that will run when their names are typed, directories that tell the names of files, etc. One can list names (and other characteristics) of files with the '1s' command in the UNIX system. Various compilers for different languages are available and numerous options for output are also usually available. More often than not, UNIX systems have a complement of "home-grown" commands in addition to those provided by the Bell System.

In Table 1.1, the prompt supplied by the system is a "%" sign. Typically, terminals connected to a UNIX system will produce a "login:" message when the break key or control D key is pressed on the terminal. After logging in, a "%" sign appears to indicate to the user that commands can be entered. With each of the commands, one must normally supply either options or file names. For example, to edit a file named junk.c, one types:

% edit junk.c or
% teco junk.c or
% ed junk.c

Since various editors are normally available (above, teco, ed, edit), the preference of the user is normally what determines which editor is used.

Chapter 5 describes the use of the UNIX system and the 'C' language in some detail with a primary orientation toward items that are useful in using UNIX-based systems in the laboratory.

1.6 REVIEW OF TEXT MATERIAL

The remaining chapters in this volume will first discuss elementary computing concepts and next describe basic instructions and addressing for the PDP-11. The UNIX system and 'C' language will be described as will 'as' and Macro, the assembly code level assemblers. The major portion of the text will focus on laboratory I/O including A/D, D/A, DI/O. Methods for acquisition of data in real time will be examined. Examples will be given in both assembly code and in 'C'. Several examples drawn from actual problems encountered in the laboratory will be described in detail.

EXERCISES

1. Write a description of the use of minicomputers in some setting with which you are familiar. You may choose an example from your own experience or read about minicomputer applications in various trade or professional journals.
2. Familiarize yourself with magazines in your library that describe minicomputer applications. For example: Electronics, Computer World, Datamation, Spectrum. Write a description of the types of articles you find in these journals.
3. If you are contemplating purchasing a computer or peripheral equipment, select several manufacturers from magazine advertisements and write to them for literature. Familiarity with manufacturer specifications and prices will help you be able to make intelligent design choices.

REFERENCE

Bell, C., and McNamara, J. E. (1978). "The PDP-8 and Other 12-bit Computers" (C. G. Bell, J. C. Mudge, and J. E. McNamara, eds.). Computer Engineering, Digital Equipment Corporation.

2

Basic Concepts

2.1 INTRODUCTION

This chapter presents fundamental information about number systems, basic computer operations, and architecture. Several basic definitions are given in Table 2.1, which the reader is encouraged to read before proceeding.

Computers typically use binary arithmetic in which a sequence of ones and zeros are used to represent each number. The correspondence between binary and decimal representation may be readily observed by comparing lines in Table 2.2. For ease of reading, most computer instructions and data can be conveniently represented in either base 8 or base 16 (octal or hexadecimal). Table 2.3 shows the relationships between base 2, 8, 10, and 16. Arithmetic in these number systems is not difficult. For example, the decimal addition

$$2_{10} + 2_{10} = 4_{10}$$

becomes

$$10_2 + 10_2 = 100_2$$

in binary. Consider octal representation. One octal number contains three bits that can represent 8_{10} binary numbers. A 12-bit number thus can be represented in two's complement form by 4 octal digits:

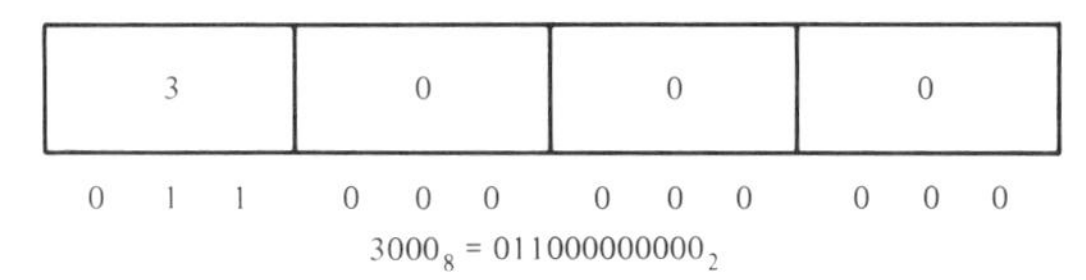

TABLE 2.1

BASIC DEFINITIONS

ASCII ASCII stands for American Standard Code for Information Interchange. This code is used to transmit serial information between various devices. The code, established by the American National Standards Institute, is commonly used for communication between computer terminals and computers and consists of 7 bits plus parity (see definition of parity below).

Bit This is the abbreviation for binary digit.

Binary digit (Bit) In binary notation the binary digit refers to either the number zero or the number one. Bit is the commonly used abbreviation for binary digit. The number of bits that are transferred in a given unit of time are typically expressed in terms of bits per second (bps). Bits per second are commonly referred to as baud.

Baud A baud is exactly the same as "bits per second" (see above) if a single event represents precisely one bit of information. A definition of baud is that it is a signaling speed that is equal to the number of single events per second or discrete conditions (e.g., zeros or ones). If, for example, one discrete signal (i.e., a zero or a one) is sent every 20 msec, the speed of transmission would be 50 baud. Common baud rates in most terminals are 300, 1200, 2400, 4800, and 9600 baud. Most terminals that are connected by telephone to a remote computer operate at 110 or 300 baud. Terminals that are connected directly to a computer via a hardwired line usually can operate at higher rates.

Byte A byte is defined as a string of binary digits usually 8 bits in length. However, for some situations, 6 bits and 9 bits may also be defined as bytes. For the PDP-11, a byte is defined as 8 bits in length.

Hardware The term hardware refers to physical equipment.

Software Software refers to programs.

Firmware Firmware is a term that refers to computer programs that are permanently fixed in a read-only memory (ROM). Firmware allows programs to remain permanently in the computer system as long as the physical hardware containing the programs is not removed.

Interface An interface is a common boundary between physical devices through which control information and data may pass. Common examples are interfaces between the computer and peripheral devices such as terminals, disks, tapes, etc.

Operating systems An operating system is a set of software that provides the communication between the operator or user of the computer system and the computer itself. Most operating systems often include a collection of utility programs that provide the user with such facilities as control of input and output, storage and management of data, program debugging, scheduling, and so forth.

Parity Parity is a term that describes the addition of noninformation bits to data in order to make the number of ones in each group of bits always either odd or even. Parity is said to be odd if the number of ones in each group of bits is always odd and even if the group has an even number of ones. This system is used for error detection.

Parallel transmission A method of data transmission in which all lines of data (zeros and ones) are transferred simultaneously between two devices.

Serial transmission A method of data transmission in which each information bit is sent sequentially over a single line rather than being sent simultaneously as in parallel transmission.

Terminal Terminals are devices that commonly provide either (1) hard copy printout of characters and a keyboard, or (2) a "glass terminal" on which characters appear in the same fashion as on a TV screen. Typical terminal rates are up to 9600 baud for glass terminals and are usually limited to 300 or 1200 baud for hard copy printout.

TABLE 2.2

CORRESPONDING DECIMAL AND BINARY NUMBERS

Decimal (base 10)	Binary (base 2)
0	00
1	01
2	10
3	11
4	100
5	101
6	110
7	111
8	1000

TABLE 2.3

THE RELATIONSHIP BETWEEN NUMBERS REPRESENTED IN BASE 2, 8, 10, AND 16

$Base_2$ binary	$Base_{10}$ decimal	$Base_8$ octal	$Base_{16}$ hexadecimal
01	1	1	1
10	2	2	2
11	3	3	3
100	4	4	4
101	5	5	5
110	6	6	6
111	7	7	7
1000	8	10	8
1001	9	11	9
1010	10	12	A
1011	11	13	B
1100	12	14	C
1101	13	15	D
1110	14	16	E
1111	15	17	F
10000	16	20	10
.	.	.	.
.	.	.	.
.	.	.	.

Conversion from one basis to another is straightforward. For example, we know that

$$132_{10} = 1 \times 10^2 + 3 \times 10^1 + 2 \times 10^0.$$

Similarly,

$$X_{10} = 8^{m-1} A_0 + 8^{m-2} A_1 + \cdots + A_{(m-1)} \tag{1}$$

for base 8. As an example, let us convert 3000_8 into decimal. Plugging into Eq. (1),

$$X = 3 \times 8^3 + 0 + 0 + 0,$$

which is equal to 1536_{10}.

Although conversions can be accomplished by formula, it is often easier to convert base 10 to base 8 using our knowledge of how binary relates to decimal. A binary–octal–decimal equivalency table (Fig. 2.1) can be used to visualize how conversions between the bases can be easily accomplished. Each bit, when set, represents a decimal number. For example, a bit in the rightmost place (called the least significant bit) represents $1 = 2^0$, where 0 = zeroth bit. For the conversion of 3000_8 to decimal, bits set in the 10th and 11th places represent 2^9 and 2^{10} contributions to the decimal number or

$$\begin{aligned} 2^{10} &= 1024 \\ 2^9 &= \underline{\ \ 512} \\ & \quad 1536_{10} \end{aligned}$$

the same answer obtained using the formula. Suppose you need to represent 100_{10} in octal. By successive subtraction you can quickly determine the octal equivalent:

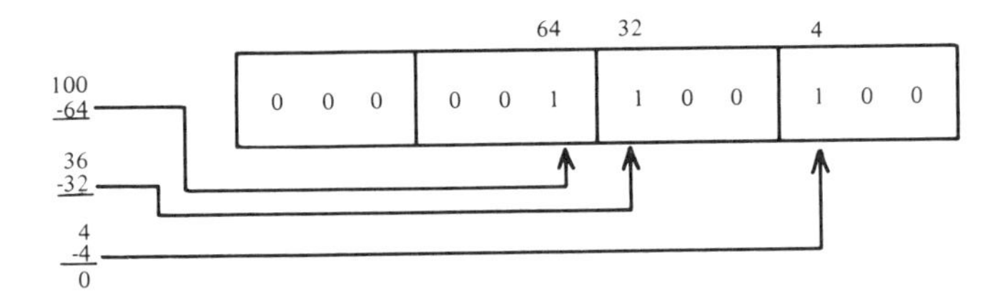

The idea is to subtract the largest possible power of 2 from the number without producing a negative result, until zero is reached. On each subtraction the bit corresponding to the value subtracted is set in the binary word. Thus

$$100_{10} = 144_8.$$

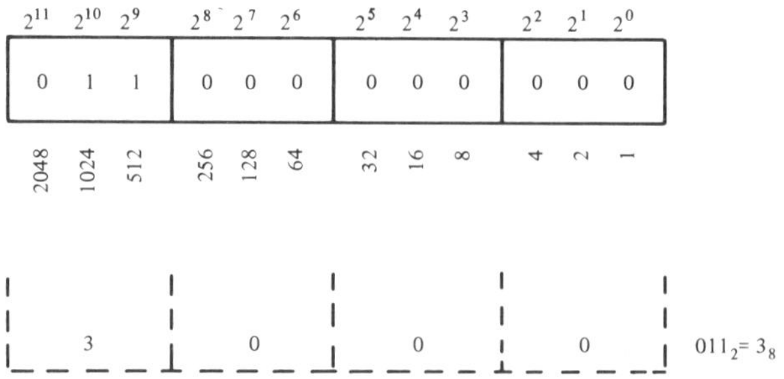

2^{11}	2^{10}	2^{9}	2^{8}	2^{7}	2^{6}	2^{5}	2^{4}	2^{3}	2^{2}	2^{1}	2^{0}
0	1	1	0	0	0	0	0	0	0	0	0
2048	1024	512	256	128	64	32	16	8	4	2	1

3 0 0 0 $011_2 = 3_8$

Fig. 2.1. Binary–octal–decimal equivalency table. Example shows that 3000_8 is equivalent to $1024_{10} + 512_{10}$.

2.2 WORDS, BITS, AND BYTES

A bit is a binary digit that can have a value of either zero or one. In electronic terms, these values can assume voltage levels such as 0 = 5 V and 1 = 0 V or vice versa. A byte is usually 8 bits (see Table 2.1) and a word is normally composed of several bytes. Different computers sometimes utilize different byte sizes. For example, on the PDP-8, a 12-bit word is used and a byte is defined as 6 bits, while on the PDP-11 a 16-bit word is employed consisting of two concatenated 8-bit bytes. Since different computers are designed with various word sizes, it is useful to consider the effect of different word sizes on arithmetic procedures. A 12-bit word can represent $4096_{10} = 2^{12}$ different values: 0 to 7777_8. A 16-bit word with 4 more bits can represent 16 times as many values as the 12-bit word or $65{,}536 = 2^{16}$. Words in computer memory can contain either instructions or data (numbers). If 8-bit bytes are employed, characters represented by 8 bits may be easily assigned to individual bytes. Negative numbers are most often represented using two's complement form in which the range of numbers that can be represented in a given word size is split into two parts to allow representation of both positive and negative numbers. Twelve bits allow signed representation of 4096_{10} values

$$3777_8 = +2047_{10}, \qquad 4000_8 = -2048_{10}$$

distributed around 0. Likewise, 16 bits provide ±32K (K = 1024) values in a single word. A single word is often used for integer arithmetic, allowing much more rapid operations than floating point arithmetic. Thus, words with lengths sufficient to handle larger numbers are advantageous in situations in which arithmetic operations are required. When only Boolean operations are necessary, shorter word lengths can be employed with no loss in speed advantage.

Both positive and negative numbers may be represented in "two's complement arithmetic." The most significant bit set in a word indicates negativity. The classic example for explaining two's complement arithmetic is the automobile odometer. If your odometer has five digits initially set to 0, and if the automobile rolls backwards, 99999 would appear, then 99998, etc. Similarly in octal, 77777 would appear for five octal digits. The octal representations corresponding to positive and negative numbers are shown in Table 2.4 for 12 and 16 bits. You can convert positive numbers into negative numbers by

(1) taking the one's complement of the number,
(2) adding one to the one's complement.

TABLE 2.4

POSITIVE AND NEGATIVE NUMBERS REPRESENTED IN TWO'S COMPLEMENT FORM FOR 12- AND 16-BIT NUMBERS

Decimal	12 bits	16 bits
32,767		077777 (maximum positive number with 16 bits)
2047	3777 (maximum positive with 12 bits)	
.	.	
.	.	
+1	1	1
0	0	0
−1	7777	177777
	7776	177776
.		.
.		.
−2048	4000 (maximum negative with 12 bits	
−32,768		100000 (maximum negative number with 16 bits)

TABLE 2.5

EXAMPLE OF CONVERSION OF 5_{10} TO OCTAL REPRESENTATION IN TWO'S COMPLEMENT FOR 12- AND 16-BIT NUMBERS

12 Bits	$5_{10} = 101_2$					
Binary	000	000	000	101		
One's complement	111	111	111	010		
Add one				+1		
Two's complement	111	111	111	011		
Octal equivalent	7	7	7	3		
16 Bits						
Binary	0	000	000	000	000	101
One's complement	1	111	111	111	111	010
Add 1						+1
Two's complement	1	111	111	111	111	011
Octal equivalent	1	7	7	7	7	3

The resulting value is a negative number in two's complement notation. The one's complement of a number is obtained by simply changing each 1 to a 0 and each 0 to a 1 in the number. For example, suppose you wish to represent −5 in two's complement with both a 12-bit and a 16-bit word. Shown above is the procedure for obtaining the correct representation. Note in Table 2.5 how octal numbers are conveniently represented in groups of three bits. After some practice it becomes quite easy to use two's complement arithmetic using octal notation.

2.3 ASCII CODES

ASCII stands for the American Standard Code for Information Interchange. Communication between users and computers and between computers employ this code in a serial fashion. The code consists of 7 units of high/low information sent in a sequence, often with a parity bit for a total of 8 bits. Parity is used to check the validity of the transmission. It is set such that the number of bits in the word becomes either odd (odd parity) or even (even parity). Consider Fig. 2.2. High/low bit streams are sent at specific predetermined rates over telephone lines, twisted pairs of dedicated lines, or other similar means. The code shown in this example is 1010101_2 or 125_8 (the code for "U"). If the transmission rate is 300 baud, 300 bits of information can be sent in a second or 30 characters (including the stop and start bits). Table 2.6 shows standard codes for characters, numbers, and control characters. There are a number of codes that per-

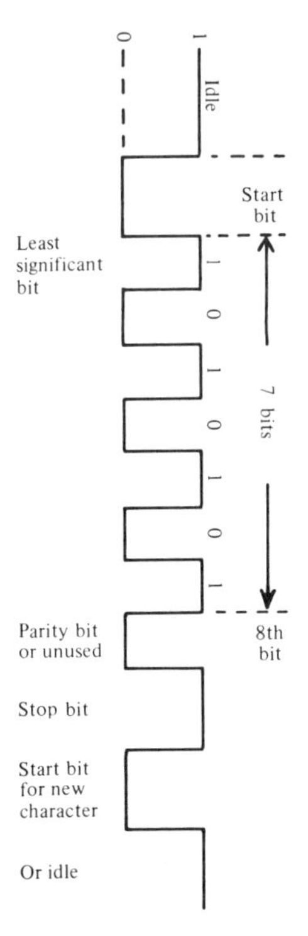

Fig. 2.2. Format for transmission of asynchronous ASCII characters.

form standard functions such as control G (^G:07), which rings the bell or beeps the buzzer on the terminal, or ^I, which produces a tab on the printer.

Source files are simply a collection of characters that are typed into the computer via the terminal. For example, to create a file using the UNIX system:

```
% make test
I This is a test CR (carriage return)
$ex$$
%
```

The codes for the characters in this file are shown in Fig. 2.3. Note that

040 = space
012 = line feed

On 16-bit machines, characters are packed 2 characters/word (1 character/byte). "This" is stored as

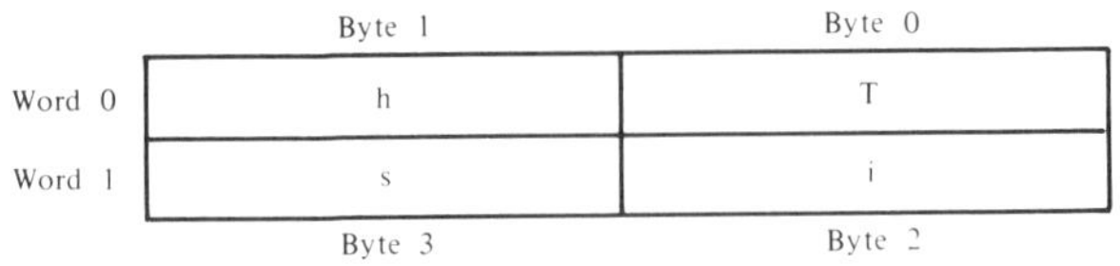

	Byte 1	Byte 0
Word 0	h	T
Word 1	s	i
	Byte 3	Byte 2

The actual words stored in memory, however, are harder to read than the above representation. Note that in this example each character is represented by 8 bits and the digits are concatenated into words as shown below. Thus, the octal word representation of the bytes containing "is" is 071551 (see Fig. 2.4).

In Fig. 2.3 the UNIX system utility 'od' was used to examine the test file. 'od' stands for octal dump. Several flags are allowable. Flags are

TABLE 2.6

STANDARD ASCII CODES FOR CHARACTERS, NUMBERS, AND CONTROL CHARACTERS

101_8 = A	141_8 = a	60 = 0	01 = ↑A (control A)
102_8 = B	142_8 = b	61 = 1	02 = ↑B (control B)
103_8 = C	143_8 = c	62 = 2	03 = ↑C (control C)
104_8 = D	144_8 = d	63 = 3	04 = ↑D (control D)
105_8 = E	145_8 = e	64 = 4	05 = ↑E (control E)
106_8 = F	146_8 = f	65 = 5	06 = ↑F (control F)
107_8 = G	147_8 = g	66 = 6	07 = ↑G (control G)
110_8 = H	150_8 = h	67 = 7	10 = ↑H (control H)
111_8 = I	151_8 = i	70 = 8	11 = ↑I (control I)
112_8 = J	152_8 = j	71 = 9	12 = ↑J (control J)
113_8 = K	153_8 = k		13 = ↑K (control K)
114_8 = L	154_8 = l		14 = ↑L (control L)
115_8 = M	155_8 = m		15 = ↑M (control M)
116_8 = N	156_8 = n		16 = ↑N (control N)
117_8 = O	157_8 = o		17 = ↑O (control O)
120_8 = P	160_8 = p		20 = ↑P (control P)
121_8 = Q	161_8 = q		21 = ↑Q (control Q)
122_8 = R	162_8 = r		22 = ↑R (control R)
123_8 = S	163_8 = s		23 = ↑S (control S)
124_8 = T	164_8 = t		24 = ↑T (control T)
125_8 = U	165_8 = u		25 = ↑U (control U)
126_8 = V	166_8 = v		26 = ↑V (control V)
127_8 = W	167_8 = w		27 = ↑W (control W)
130_8 = X	170_8 = x		30 = ↑X (control X)
131_8 = Y	171_8 = y		31 = ↑Y (control Y)
132_8 = Z	172_8 = z		32 = ↑Z (control Z)

```
% make test
*2ET-1^X1ED$$
*EWtest$$
*iThis is a test
$$
*ex$$
% od test
0000000 064124 071551 064440 020163 020141 062564 072163 000012
0000017
% od -b test
0000000 124 150 151 163 040 151 163 040 141 040 164 145 163 164 012 000
0000017
% od -c test
0000000  T  h  i  s     i  s     a     t  e  s  t \n \0
0000017
% od -o test
0000000 064124 071551 064440 020163 020141 062564 072163 000012
0000017
% od -n test
0000000 064124 071551 064440 020163 020141 062564 072163 000012
0000017
% od -d test
0000000 26708 29545 26912 08307 08289 25972 29811 00010
0000017
```

Fig. 2.3. Example of using the UNIX octal dump (od) program to print the contents of a file in decimal, octal, ASCII, etc.

commands to the program that cause the data to be printed differently. In this case

- −b: print in bytes
- −c: print in characters
- −o: print in octal words
- −h: print in hexadecimal
- −d: print in decimal

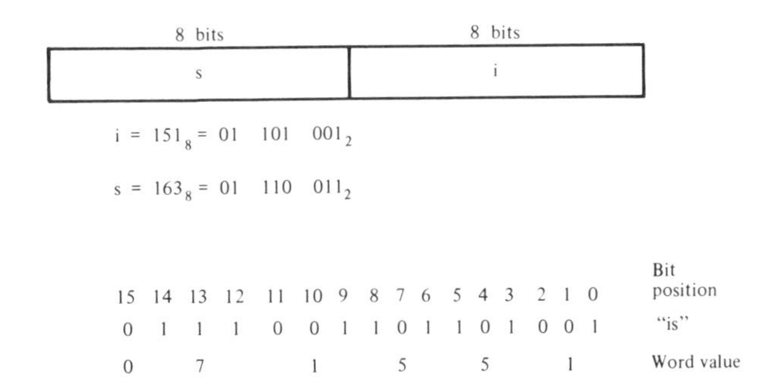

Fig. 2.4. Octal word representation of "is".

In the left column the address is printed. The first row contains addresses 0 through 16_8 and 17 begins the second row. The % sign is the UNIX system prompt. At the top of the figure, the actual commands used to enter the characters into the file are reproduced.

Bytes may be packed differently for different word lengths. For example, one way of packing bytes in a 12-bit word is

11 — 8	7 — 0
Char 2	Character 1
Char 2	Character 3

where the second character is packed in the upper 4 bits of two consecutive words.

2.4 MULTIPLICATION EXAMPLE

The following example is presented as a way of conceptually introducing the ideas associated with producing algorithms in assembly code. Multiplication of two numbers provides a simple example that introduces the use of registers, rotation, program flow, etc.

Multiplication can be accomplished by a series of additions. For example,

$$\begin{array}{r} 11_{10} \\ \underline{12_{10}} \\ 22 \\ \underline{110} \\ 132_{10} \end{array} = [11 + 11] + [11 \text{ shifted by } 10 = 110] = 132$$

Since adding and shifting are two common operations on minicomputers, a simple algorithm can be written to multiply two numbers. The following example will multiply only positive numbers and is not intended to be a particularly efficient or general method. Codes that will instruct the computer to conduct the desired operations (e.g., add) called assembly code, will be examined later. For now, we shall explain the operations in descriptive terms such as add and rotate (shift).

Consider multiplying $5_{10} \times 3_{10}$:

$$\begin{array}{r} 011_2 \\ \underline{101_2} \\ 011 \\ \underline{011} \\ 01111 \end{array} = 17_8 = 15_{10}$$

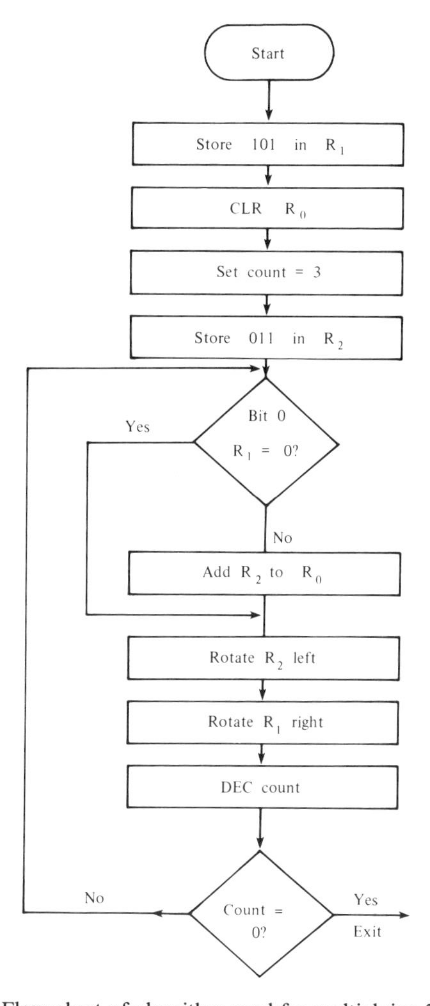

Fig. 2.5. Flow chart of algorithm used for multiplying $3_{10} \times 5_{10}$.

Suppose we have three general registers, R0, R1, and R2, which can be used for adding or subtracting numbers or can be used to rotate the bits left or right. Figure 2.5 shows the flow chart for multiplying 3×5 and Table 2.7 shows the values in each register at each iteration through the main loop in the program. Also shown are the contents of a memory location named count. The procedure for multiplying 3×5 is

1. $101_2 = (5)$ is stored in R1, $011_2 = (3)$ is stored in R2, R0 is set $= 0$. Count is set $= 3 =$ number of iterations.

TABLE 2.7

STEPS IN THE MULTIPLICATION OF $3_{10} \times 5_{10}$

Step	Register contents	
1.	011	R_2
	101	R_1
	011	R_0
2.	0110	R_2
	10	R_1
	011	R_0
3.	01100	R_2
	1	R_1
	01111	R_0

Step	Count	R_0	R_1	R_2
1	3	011	101	011
2	2	011	10	0110
3	1	01111	1	1100

2. Each bit in R1 (the multiplier) is examined. If the bit = 1, the multiplicand (3) = R2 is added to the register R0. On each iteration, R2 is shifted left once, but only added to R0 when a bit in R1 is = 1.

The major ideas that should be learned from this example are (1) that general purpose registers are used in arithmetic operations in minicomputers and (2) that algorithms for conducting various operations (such as multiplying two numbers) can be implemented, using instructions (e.g., add and rotate) that are available for use with general purpose registers. In subsequent sections we shall explore in detail the various types of instructions that are available.

2.5 MINICOMPUTER ARCHITECTURE

The term architecture, when used in reference to computers, generally refers to the set of resources available on a computer including the central processing unit, instructions, memory, registers, etc. Minicomputers, microprocessors, and larger computers share many fundamental concepts. Although there are many variations in instructions and addressing, a common conceptual thread can be followed that will allow the reader to understand almost any machine. For simplicity, the more detailed discussions that follow will describe the PDP-8 and PDP-11. The PDP-8 will not be used later in this text, but is useful here since it has an extremely simple instruction set and addressing scheme.

2.5.1 PDP-8

The PDP-8 has eight distinct instructions that are each assigned a specific code: 0–7. At the leftmost part (most significant) of the 12-bit word, 3 bits are allocated for an operation code (op code) as shown in Fig. 2.6. The PDP-8 has an accumulator (AC) (a general purpose register) and a series of instructions (also shown in Fig. 2.6), which operate on the AC and on memory locations. Since it is beyond the scope of this text to discuss PDP-8 programming, the reader is referred to "Introduction to Programming" by Digital Equipment Corporation. Nevertheless, it is instructive to consider what can be done with the remaining 9 bits of a computer word after 3 bits are used for the op code. Suppose we designed the machine so that we could address 2^9 bits or 512 words and that all 9 bits were used to contain an address. Instructions would be quite simple. As an example, let us add the contents of location 2 to the contents of location 3 and put the result in location 511:

Symbolic instruction	Code	
TAD 2	1002	/add what is in location two to AC
TAD 3	1003	/add what is in 3 to AC
DCA 511	3777	/put in 511, $777_8 = 511_{10}$

Memory would appear as shown in Fig. 2.7 if the values 10 and 20 were in locations 2 and 3 before the program was executed. The reader may already see the drawback to this scheme—only 512 words can be used. If we made the word 13 bits long, then 2^{10} or 1024 locations could be addressed. The notion, however, of continuing to expand the word size in this way is impractical. Designers actually employed a much better method for addressing that uses a technique known as *indirect or deferred addressing*. For example, the instruction format might look like

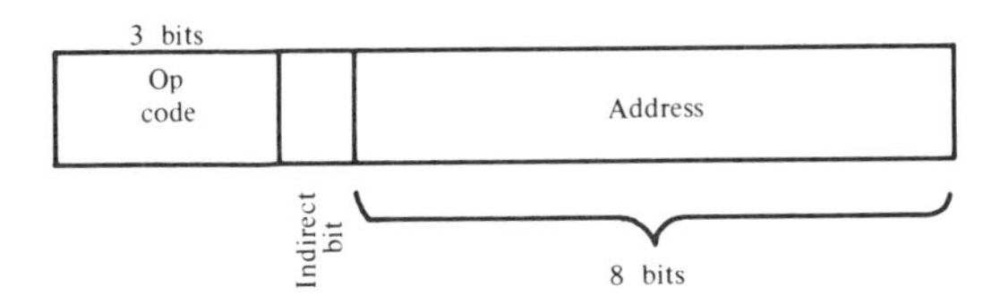

11	10	9	8	7	6	5	4	3	2	1	0
Op code					Ad	dress			Field		

Op code		Mnemonic		Description
0	=	AND	-	And
1	=	TAD	-	Two's complement add
2	=	ISZ	-	Increment and skip
3	=	DCA	-	Deposit in accumulator
4	=	JMS	-	Jump to a subroutine
5	=	JMP	-	Jump
6	=	IOT	-	Input output transfer
7	=	Operate	-	Operate (a series of basic instructions, e.g., rotate , clear , etc.)

Fig. 2.6. Instruction format and instructions for the PDP-8.

When bit 8 is set, the address field is used to specify the address of a location that contains the address of the desired number. Suppose we code the symbolic instruction for indirect addressing as: TAD I 2 where the symbol I sets the indirect bit. "TAD I 2" means use the value in location 2 *to point to* the address with the value we are to add to the AC. For the numbers in memory shown in Fig. 2.8, "TAD I 2" would put 10 in the AC, whereas "TAD 2" would put 6000 in the AC.

The PDP-8 actually only uses a 7-bit address field as shown in Fig. 2.9. Another bit is added that, when not set, indicates that the address references the first 128 words in memory (called page 0 on the PDP-8). When

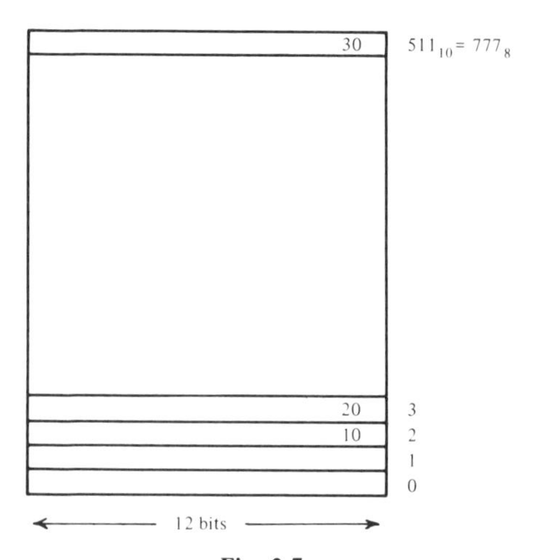

Fig. 2.7.

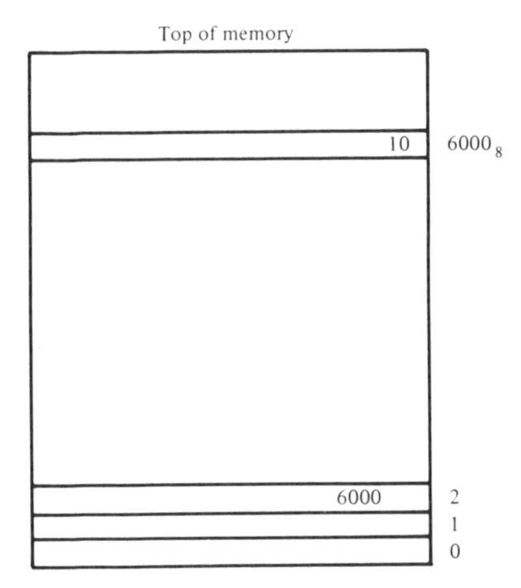

Fig. 2.8.

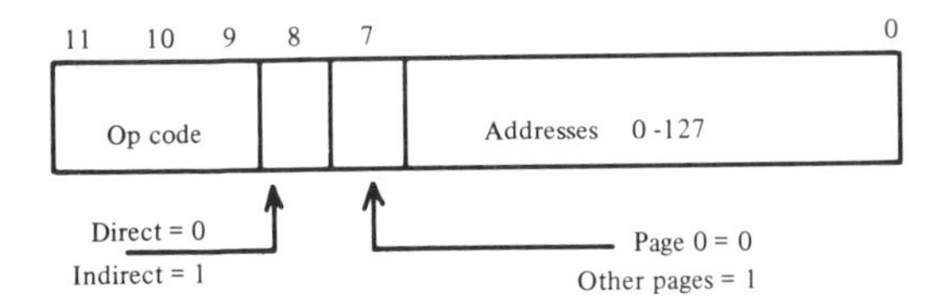

Fig. 2.9. Details of PDP-8 instruction format.

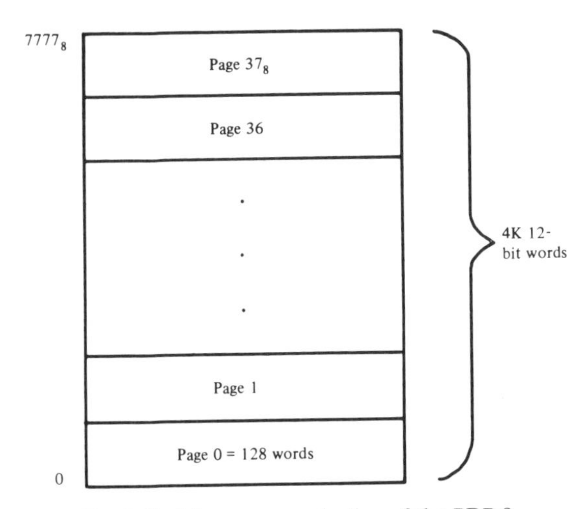

Fig. 2.10. Memory organization of the PDP-8.

this bit in an instruction is set, the address referred to is in the same 128 word segment of memory as the instruction. Memory is allocated into 128 word (12-bit) blocks or chunks of memory called pages that make up a "field" of 4K words (see Fig. 2.10). A PDP-8 may contain up to eight fields for a total of 32K words of memory.

The basic observation is that PDP-8 code is simple but due to a short word size must break up the program into many small pieces. Other machines use different methods. We shall next describe in detail a different architectural concept that is used for the PDP-11.

2.5.2 PDP-11 Fundamentals

The PDP-11 was introduced by Digital Equipment Corporation about 1970 and enjoyed considerable success in the marketplace during the 1970s. Although many models have been produced, the fundamental architecture has retained considerable similarity between older and newer models. Since this book is predominantly about the use of PDP-11s in the laboratory, we shall examine the PDP-11 addressing modes and instruction set in detail.

One may represent the block structure of the PDP-11 as shown in Fig. 2.11. The processor has eight general purpose registers. R7 is designated as the program counter (PC). The PC contains the address following the instruction in memory, which is being executed. R6 is used for the address of the stack pointer, which points to an area reserved for temporary storage of information. The processor status word (PSW) continuously changes while a program is running, indicating the status of the processor. Various bits in the PSW, called flags or condition codes, indicate whether the previous operation resulted in a carry (C), overflow (V), zero result (Z), or negative result (N). Whenever an instruction is executed, any of these bits may be set, cleared or left unchanged. For example, one instruction is the compare (cmp) instruction. To compare X and Y, one writes

cmp X,Y

If comparing X and Y by subtracting Y from X results in a negative value, the N bit is set. One can test values and check bits in the PSW, e.g.,

tst location

If the value in the address labeled by "location" is zero, the Z bit in the PSW is set. This information can then be tested by a branch instruction to control program flow. A "T" bit is available to indicate, if the bit is set,

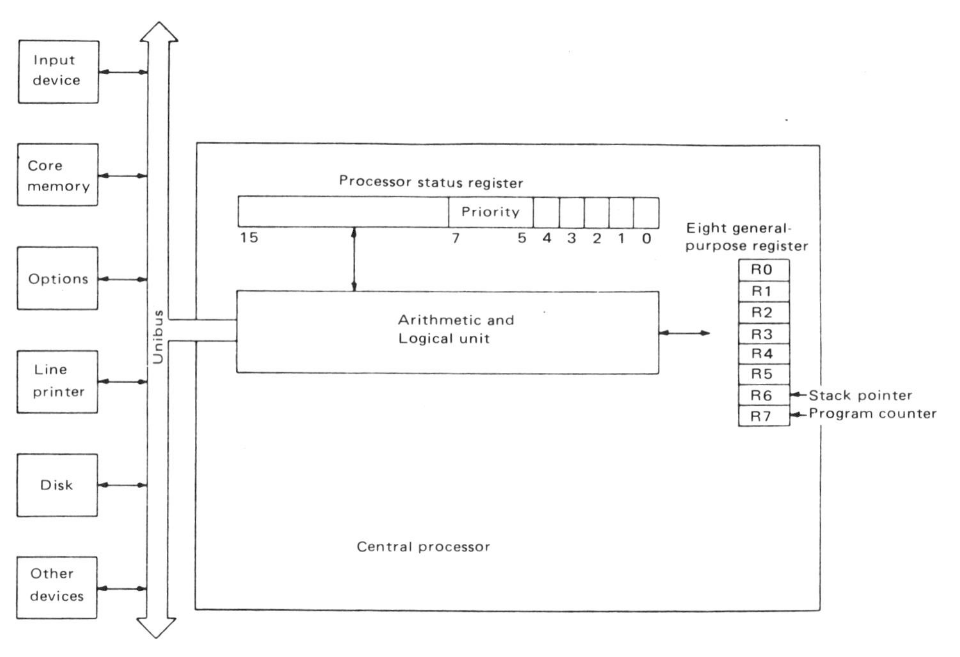

Fig. 2.11. PDP-11 block structure reproduced from the "PDP-11 Processor Handbook."

that a jump to a specific location (or "trap") should be made after each instruction. Three bits (bits 5, 6, and 7) are available for priority specification (see Chapters 3 and 8).

Memory is organized by bytes and words as shown in Fig. 2.12. The number of words and bytes that can be used with a PDP-11 can be computed by considering a 16-bit word (see Fig. 2.13). A bit set in any position

High byte	Low byte	Word location
Byte 1	Byte 0	0
Byte 3	Byte 2	2
Byte 5	Byte 4	4

Fig. 2.12. Organization of memory on the PDP-11.

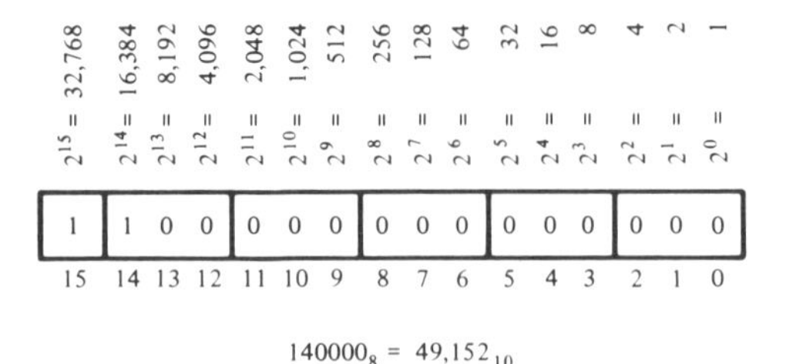

Fig. 2.13. Sixteen-bit word.

N represents the decimal number 2^N. Thus, a bit set in bit 15 (100000_8) represents an unsigned number of 2^{15} or 32,768. Setting bit 14 in addition (140000_8) represents $2^{15} + 2^{14} = 32{,}768 + 16{,}384$. Continuing similarly, if we set all bits (177777_8), we can represent 65,535 or 2^{16}-1. Since bytes receive individual numbers as addresses on the PDP-11, 65,536 bytes or 32,768 words can be represented with a 16-bit word length. The correspondence between byte locations and number of words in memory is shown in Table 2.8. Thus a 28K word machine (a common configuration) would have byte addresses from 0 to 157,777 (16K + 8K + 4K words = 100,000 + 40,000 + 20,000 = 160,000 − 1 = 157,777 bytes).

On small PDP-11s, without memory management, only 32K words can be accessed. The locations 0–400 (base 8) are used for interrupt vector locations to various devices and the memory locations between 28 and 32K are used for device registers and are not normally available to the user. Additional information on this arrangement is given in the next chapter.

TABLE 2.8

CORRESPONDING ADDRESSES AND NUMBER OF BYTES AND WORDS IN A 28K-WORD MEMORY

$Bytes_8$	$Bytes_{10}$	$Words_{10}$
2,000	1024	512
4,000	2048	1K
10,000	4096	2K
20,000	8K	4K
40,000	16K	8K
100,000	32K	16K
157,776	56K-2	28K-1

2.6 PDP-11 INSTRUCTIONS

The PDP-11 is a variable instruction length processor, using either 1, 2, or 3 words for a single instruction. These instructions fall in three groups as shown in Fig. 2.14. "Instruction" refers to a specific command and "operand" refers to that which is operated upon. The operand may be a register or a register can refer directly or indirectly to another location. Also, the address of the operand may be built from the addition of two addresses. In some cases, two words are needed for a single-operand instruction and, in the double-operand case, up to three words may be required. The entire set of commands is shown in Table 2.9 and each group is explained in more detail in what follows. This table is reprinted from an LSI-11 reference card supplied by Digital Equipment Corporation.

All addressing is accomplished via the registers. Although eight registers are available, only six are generally employed by the user. R7 is used as the program counter (PC) and always contains the address of the instruction after the instruction that is currently being executed. Addressing relative to register 7 is very common. In this mode, the address or addresses following the instruction contain offsets from the PC that can be used to form the address of the operand. This concept is explained in detail in Section 2.6.3. R6 is defined as the stack pointer (SP) and contains the address of an area in memory that is used for temporary storage and other uses (see subroutine section in this chapter).

Instructions are formatted differently depending on whether they are in the operate group, single-operand, or double-operand group. Operate refers to a set of instructions that requires no operand (for example, HALT). An octal code is assigned for each operate instruction. HALT is

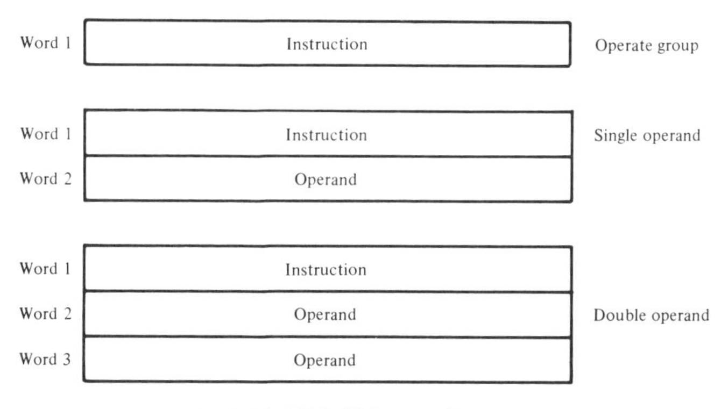

Fig. 2.14. PDP-11 instruction groups.

TABLE 2.9

LSI-11, PDP 11/03 REFERENCE CARD[a]

LSI-11, PDP-11/03 Reference Card

OCTOBER 1976

WORD FORMAT

15 14 12 11 9 8 6 5 3 2 0

BINARY-OCTAL REPRESENTATION

MODE | R

Mode	Name	Symbolic	Description
0	register	R	(R) is operand [ex. R2=%2]
1	register deferred	(R)	(R) is address
2	auto-increment	(R)+	(R) is adrs; (R) +(1 or 2)
3	auto-incr deferred	@(R)+	(R) is adrs of adrs; (R) + 2
4	auto-decrement	−(R)	(R) − (1 or 2); is adrs
5	auto-decr deferred	@−(R)	(R) − 2; (R) is adrs of adrs
6	index	X(R)	(R) + X is adrs
7	index deferred	@X(R)	(R) + X is adrs of adrs

PROGRAM COUNTER ADDRESSING Reg = 7

MODE | 7

2	immediate	#n	operand n follows instr
3	absolute	@#A	address A follows instr
6	relative	A	instr adrs + 4 + X is adrs
7	relative deferred	@A	instr adrs + 4 + X is adrs of adrs

LEGEND

Op Codes

■ = 0 for word/1 for byte
SS = source field (6 bits)
DD = destination field (6 bits)
R = gen register (3 bits), 0 to 7
XXX = offset (8 bits), +127 to −128
N = number (3 bits)
NN = number (6 bits)

Boolean

∧ = AND
∨ = inclusive OR
∀ = exclusive OR
∽ = NOT

Operations

( ) = contents of
s = contents of source
d = contents of destination
r = contents of register
← = becomes
X = relative address
% = register definition

Condition Codes

* = conditionally set/cleared
— = not affected
0 = cleared
1 = set

digital

DIGITAL EQUIPMENT CORPORATION
Component Group Headquarters: 1 Iron Way
Marlborough, Mass. 01752, Telephone: (617) 481-7400

TABLE 2.9 (continued)

SINGLE OPERAND: OPR dst

15 … 6	5 … 0
OP CODE	SS OR DD

Mnemonic	Op Code	Instruction	dst Result	N	Z	V	C
General							
CLR(B)	■ 050DD	clear	0	0	1	0	0
COM(B)	■ 051DD	complement (1's)	~d	*	*	0	1
INC(B)	■ 052DD	increment	d + 1	*	*	*	–
DEC(B)	■ 053DD	decrement	d – 1	*	*	*	–
NEG(B)	■ 054DD	negate (2's compl)	–d	*	*	*	*
TST(B)	■ 057DD	test	d	*	*	0	0
Rotate & Shift							
ROR(B)	■ 060DD	rotate right	→ C, d	*	*	*	*
ROL(B)	■ 061DD	rotate left	C, d ←	*	*	*	*
ASR(B)	■ 062DD	arith shift right	d/2	*	*	*	*
ASL(B)	■ 063DD	arith shift left	2d	*	*	*	*
SWAB	0003DD	swap bytes		*	*	0	0
Multiple Precision							
ADC(B)	■ 055DD	add carry	d + C	*	*	*	*
SBC(B)	■ 056DD	subtract carry	d – C	*	*	*	*
SXT	0067DD	sign extend	0 or –1	–	*	0	–
Processor Status (PS) Operators							
MFPS	1067DD	move byte from PS	d ← PS	*	*	0	–
MTPS	1064SS	move byte to PS	PS ← s	*	*	*	*

DOUBLE OPERAND: OPR src, dst OPR src, R or OPR R, dst

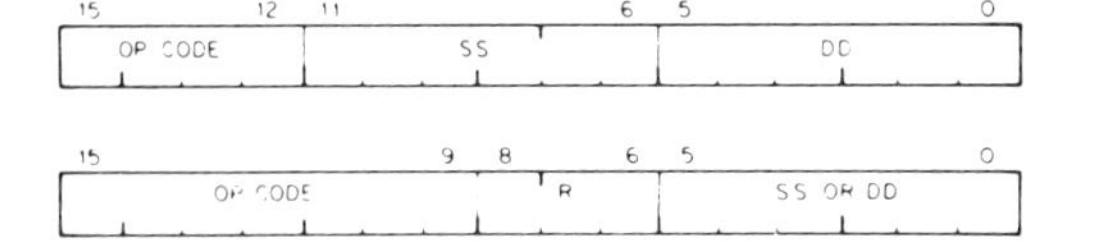

Mnemonic	Op Code	Instruction	Operation	N	Z	V	C
General							
MOV(B)	■ 1SSDD	move	d ← s	*	*	0	–
CMP(B)	■ 2SSDD	compare	s – d	*	*	*	*
ADD	06SSDD	add	d ← s + d	*	*	*	*
SUB	16SSDD	subtract	d ← d – s	*	*	*	*
Logical							
BIT(B)	■ 3SSDD	bit test (AND)	s ∧ d	*	*	0	–
BIC(B)	■ 4SSDD	bit clear	d ← (~s) ∧ d	*	*	0	–
BIS(B)	■ 5SSDD	bit set (OR)	d ← s ∨ d	*	*	0	–
XOR	074RDD	exclusive OR	d ← r ∨ d	*	*	0	–

TABLE 2.9 (continued)

Optional EIS

MUL	070RSS	multiply	r ← r x s	*	*	0	*
DIV	071RSS	divide	r ← r/s	*	*	*	*
ASH	072RSS	shift arithmetically		*	*	*	*
ASHC	073RSS	arith shift combined		*	*	*	*

Optional FIS

FADD	07500R	floating add		*	*	0	0
FSUB	07501R	floating subtract		*	*	0	0
FMUL	07502R	floating multiply		*	*	0	0
FDIV	07503R	floating divide		*	*	0	0

BRANCH: B - - location

If condition is satisfied:
Branch to location,
New PC ← Updated PC + (2 x offset)
(Updated PC = adrs of br instr + 2)

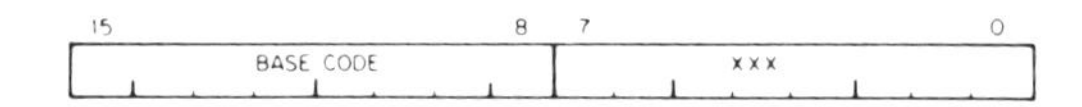

Op Code = Base Code + XXX

Mnemonic	Base Code	Instruction	Branch Condition	
Branches				
BR	000400	branch (unconditional)	(always)	
BNE	001000	br if not equal (to 0)	≠ 0	Z = 0
BEQ	001400	br if equal (to 0)	= 0	Z = 1
BPL	100000	branch if plus	+	N = 0
BMI	100400	branch if minus	—	N = 1
BVC	102000	br if overflow is clear		V = 0
BVS	102400	br if overflow is set		V = 1
BCC	103000	br if carry is clear		C = 0
BCS	103400	br if carry is set		C = 1
Signed Conditional Branches				
BGE	002000	br if greater or equal (to 0)	≥0	N ⊻ V = 0
BLT	002400	br if less than (0)	<0	N ⊻ V = 1
BGT	003000	br if greater than (0)	>0	Z v (N ⊻ V) = 0
BLE	003400	br if less or equal (to 0)	≤0	Z v (N ⊻ V) = 1
Unsigned Conditional Branches				
BHI	101000	branch if higher	>	C v Z = 0
BLOS	101400	branch if lower or same	≤	C v Z = 1
BHIS	103000	branch if higher or same	≥	C = 0
BLO	103400	branch if lower	<	C = 1

TABLE 2.9 (continued)

JUMP & SUBROUTINE

Mnemonic	Op Code	Instruction	Notes
JMP	0001DD	jump	PC ← dst
JSR	004RDD	jump to subroutine	use same R
RTS	00020R	return from subroutine	use same R
MARK	0064NN	mark	aid in subr return
SOB	077RNN	subtract 1 & br (if ≠ 0)	(R) — 1, then if (R) ≠ 0: PC ← Updated PC — (2 x NN)

TRAP & INTERRUPT:

Mnemonic	Op Code	Instruction	Notes
EMT	104000 to 104377	emulator trap (not for general use)	PC at 30, PS at 32
TRAP	104400 to 104777	trap	PC at 34, PS at 36
BPT	000003	breakpoint trap	PC at 14, PS at 16
IOT	000004	input/output trap	PC at 20, PS at 22
RTI	000002	return from interrupt	
RTT	000006	return from interrupt	inhibit T bit trap

MISCELLANEOUS:

Mnemonic	Op Code	Instruction
HALT	000000	halt
WAIT	000001	wait for interrupt
RESET	000005	reset external bus
NOP	000240	(no operation)

CONDITION CODE OPERATORS:

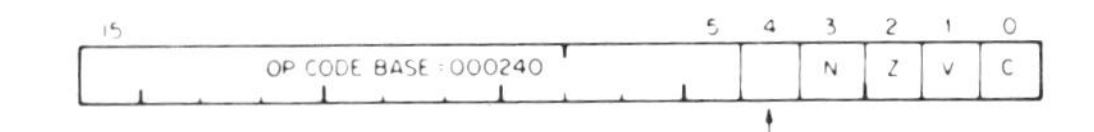

0 = CLEAR SELECTED COND. CODE BITS
1 = SET SELECTED COND. CODE BITS

Mnemonic	Op Code	Instruction	N	Z	V	C
CLC	000241	clear C	–	–	–	0
CLV	000242	clear V	–	–	0	–
CLZ	000244	clear Z	–	0	–	–
CLN	000250	clear N	0	–	–	–
CCC	000257	clear all cc bits	0	0	0	0
SEC	000261	set C	–	–	–	1
SEV	000262	set V	–	–	1	–
SEZ	000264	set Z	–	1	–	–
SEN	000270	set N	1	–	–	–
SCC	000277	set all cc bits	1	1	1	1

TABLE 2.9 (continued)

NUMERICAL OP CODE LIST

OP Code	Mnemonic
00 00 00	HALT
00 00 01	WAIT
00 00 02	RTI
00 00 03	BPT
00 00 04	IOT
00 00 05	RESET
00 00 06	RTT
00 00 07 – 00 00 77	(unused)
00 01 DD	JMP
00 02 0R	RTS
00 02 10 ↕ 00 02 27	(reserved)
00 02 40	NOP
00 02 41 ↕ 00 02 77	cond codes
00 03 DD	SWAB
00 04 XXX	BR
00 10 XXX	BNE
00 14 XXX	BEQ
00 20 XXX	BGE
00 24 XXX	BLT
00 30 XXX	BGT
00 34 XXX	BLE
00 4R DD	JSR
00 50 DD	CLR
00 51 DD	COM
00 52 DD	INC
00 53 DD	DEC
00 54 DD	NEG
00 55 DD	ADC
00 56 DD	SBC
00 57 DD	TST

OP Code	Mnemonic
00 60 DD	ROR
00 61 DD	ROL
00 62 DD	ASR
00 63 DD	ASL
00 64 NN	MARK
00 67 DD	SXT
00 70 00 ↕ 00 77 77	(unused)
01 SS DD	MOV
02 SS DD	CMP
03 SS DD	BIT
04 SS DD	BIC
05 SS DD	BIS
06 SS DD	ADD
07 0R SS	MUL
07 1R SS	DIV
07 2R SS	ASH
07 3R SS	ASHC
07 4R DD	XOR
07 50 0R	FADD
07 50 1R	FSUB
07 50 2R	FMUL
07 50 3R	FDIV
07 50 40 ↕ 07 67 77	(unused)
07 7R NN	SOB
10 00 XXX	BPL
10 04 XXX	BMI
10 10 XXX	BHI
10 14 XXX	BLOS
10 20 XXX	BVC
10 24 XXX	BVS
10 30 XXX	BCC, BHIS
10 34 XXX	BCS, BLO

OP Code	Mnemonic
10 40 00 ↕ 10 43 77	EMT
10 44 00 ↕ 10 47 77	TRAP
10 50 DD	CLRB
10 51 DD	COMB
10 52 DD	INCB
10 53 DD	DECB
10 54 DD	NEGB
10 55 DD	ADCB
10 56 DD	SBCB
10 57 DD	TSTB
10 60 DD	RORB
10 61 DD	ROLB
10 62 DD	ASRB
10 63 DD	ASLB
10 64 SS	MTPS
10 67 DD	MFPS
11 SS DD	MOVB
12 SS DD	CMPB
13 SS DD	BITB
14 SS DD	BICB
15 SS DD	BISB
16 SS DD	SUB
17 00 00 ↕ 17 77 77	RE-SERVED

RESERVED TRAP AND INTERRUPT VECTORS

Vector	Use
000	(Reserved)
004	Bus Timeout and Illegal Instructions (eg. JMP R0) (Odd Address and Stack Overflow Traps Not Implemented on LSI-11)
010	Illegal and Reserved Instruction
014	BPT Instruction and T Bit
020	IOT Instruction
024	Power Fail
030	EMT Instruction
034	TRAP Instruction
060	Console Input Device
064	Console Output Device
100	External Event Line Interrupt
200	LAV11
244	FIS (Optional)
264	RXV11
300	Floating Vectors start here

TABLE 2.9 (continued)

7-BIT ASCII CODE

Octal Code	Char	Octal Code	Char	Octal Code	Char	Octal Code	Char
000	NUL	040	SP	100	@	140	\
001	SOH	041	!	101	A	141	a
002	STX	042	"	102	B	142	b
003	ETX	043	#	103	C	143	c
004	EOT	044	$	104	D	144	d
005	ENQ	054	%	105	E	145	e
006	ACK	046	&	106	F	146	f
007	BEL	047	'	107	G	147	g
010	BS	050	(	110	H	150	h
011	HT	051	)	111	I	151	i
012	LF	052	*	112	J	152	j
013	VT	053	+	113	K	153	k
014	FF	054	,	114	L	154	l
015	CR	055	-	115	M	155	m
016	SO	056	.	116	N	156	n
017	SI	057	/	117	O	157	o
020	DLE	060	0	120	P	160	p
021	DC1	061	1	121	Q	161	q
022	DC2	062	2	122	R	162	r
023	DC3	063	3	123	S	163	s
024	DC4	056	4	124	T	164	t
025	NAK	065	5	125	U	165	u
026	SYN	066	6	126	V	166	v
027	ETB	067	7	127	W	167	w
030	CAN	070	8	130	X	170	x
031	EM	071	9	131	Y	171	y
032	SUB	072	:	132	Z	172	z
033	ESC	073	;	133	[	173	{
034	FS	074	<	134	\	174	|
035	GS	075	=	135	] or ↑	175	}
036	RS	076	>	136	∧	176	~
037	US	077	?	137	— or ←	177	DEL

PROCESSOR STATUS WORD

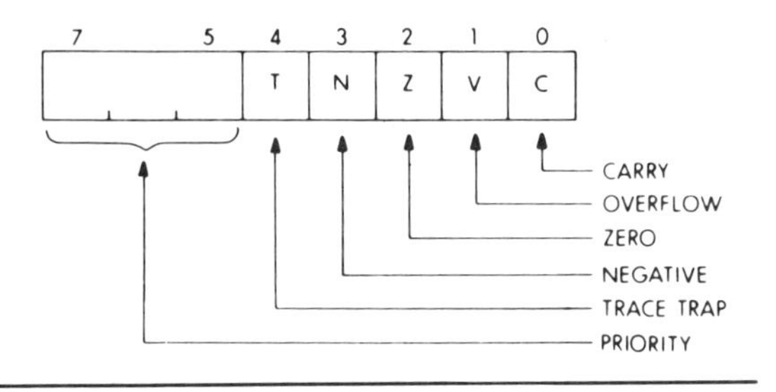

defined as 000000. Thus, whenever a 0 is executed by the machine, the machine will halt.

2.6.1 Single Operand

Single-operand instructions operate on one operand. The instruction format is

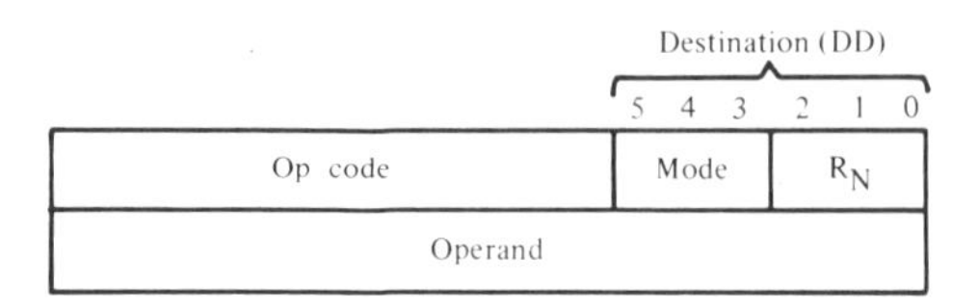

where R_N refers to a register and mode refers to one of eight modes. The word following the instruction may be used in some modes to form the operand.

An example of a single-operand instruction that requires only one word is

CLR R4

If we wish to clear R4, the symbolic instruction above may be issued. Examine Table 2.9 to determine what the actual octal code will be that the machine will use to clear R4. The code found in the table is 050 (CLR). The destination (DD) is R4, thus $R_N = 4$ and the mode is register mode since we are referring to an operation on a register. Thus mode = 0 and the decoded instruction is 05004. Note the format of the mode:

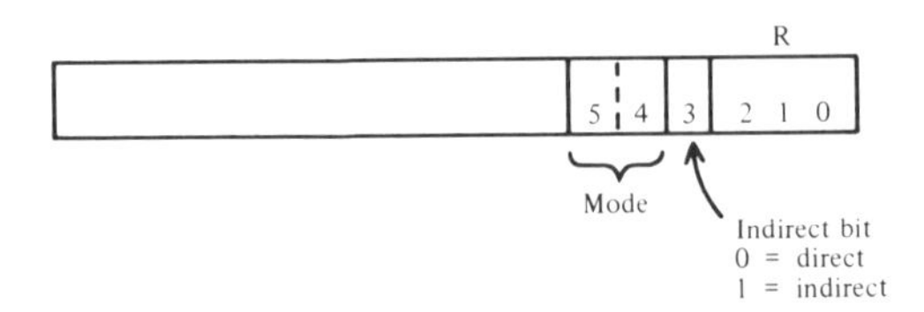

For both direct and indirect addressing, bits 4 and 5 are

Mode	Mode name
00	Register
01	Autoincrement
10	Autodecrement
11	Indexed

Therefore, for direct and indirect addressing, by adding a zero or a 1 to the end of the string, we have for the concatenation of the mode and indirect bits:

Direct	Indirect
000	001
010	011
100	101
110	111

Table 2.10 summarizes the modes in tabular form including symbols for both 'as' and Macro.

Consider the following examples of use of different methods of addressing:

1. Direct Addressing

(a) Register mode (mode 000) symbol: r or R, where $0 \leq r \leq 7$.

(Note that in each part of this chapter subsection that describes the types of addressing available, the symbols for both 'as' and Macro are given. In each case the symbol for 'as' is followed by the symbol for Macro.)

symbolic code	general machine code	specific machine code
inc r3	052DD	05203

Above, the value in r3 is incremented.

If the code being executed is

Memory loc	Instruction
.	.
.	.
1000	05003
1002	05203
1004	00000
.	.
.	.
.	.

r3 will be cleared (050), then incremented (052), and the machine halted (0). An example of running this code on an 11/03, using ODT microcode, will be given in Section 2.7.

(b) Autoincrement (mode 010) and decrement (mode 100) modes:

Symbols: Autoincrement: (r)+ or (R)+
Autodecrement: −(r) or −(R)

TABLE 2.10

PDP-11 ADDRESSING MODES

Mode	Name	'as' symbol	Macro symbol
	1. Direct addressing		
0	Register	r	R
2	Autoincrement	(r)+	(R)+
4	Autodecrement	−(r)	−(R)
6	Index	x(r)	x(R)
	2. Deferred or indirect addressing		
1	Register indirect	(r)	(R)
3	Autoincrement indirect	*(r)+	@(R)+
5	Autodecrement indirect	*−(r)	@−(R)
7	Index indirect	*x(r)	@x(R)
	3. Use of R_7 as a general register		
2	Immediate	$n	#n
3	Absolute	*$A	@#A
6	Relative	A	A
7	Relative indirect	*A	@A

Both autoincrement and autodecrement instructions are provided in the instruction set. For autoincrement, the address in the register is used to refer to the operand and is incremented after use. The increment will be either one or two depending on whether bytes or words are referenced. That is, com (r5)+ increments r5 by two and comb (r5)+ will increment r5 by one. This instruction decodes as

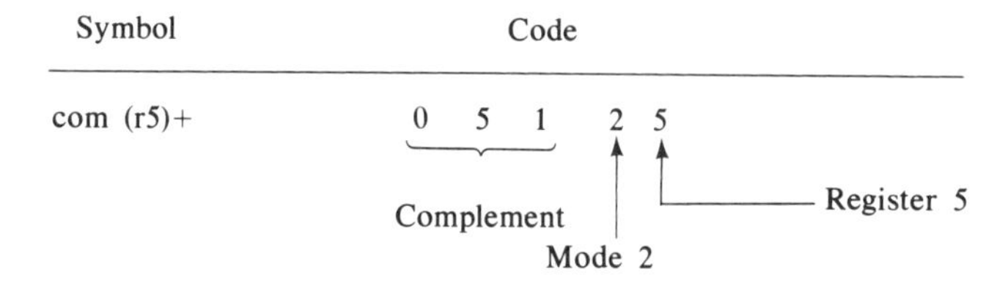

In this example, r5 is used as the address of the operand. Suppose, for example that "r5" is incremented by 2 (!!) to point to the next *word* (!!) location *after* it is used in the instruction (see Fig. 2.15). Use of autoindexing in this manner is useful for performing an operation on a sequential list. Autodecrement is conceptually the same as autoincrement except a decrement occurs first. That is, prior to using the value in the register, the contents of the register are decremented by one (byte) or two (word). For

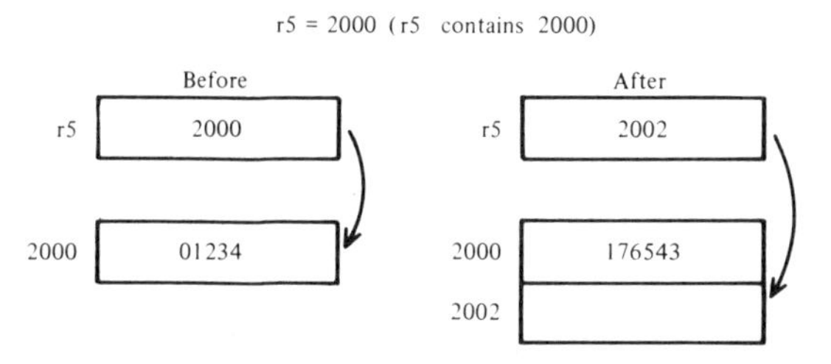

Fig. 2.15. Example of com (r5)+.

an example, see Fig. 2.16. It is important to observe that $2000_8 - 2_8 = 1776_8$ and that 1776_8 is used as the address of the operand in this example.

(c) Indexed (mode 110) mode. Symbol: x(r) or X(R). Indexing allows access of an address by adding a value x and the contents of a register to form the address to be accessed. An example of such code is

Symbol	Code
com x(r5)	0 5 1 6 5
	0 5 1 11 0 101

(11 = mode 11; 0 = Direct; 101 = r5)

Suppose x is equal to 200 and r5 contains 1000, com x(r5) will complement what is in 1000 + 200 = 1200. That is,

Before	After
x = 200	x = 200
r5 = 1000	r5 = 1000
1200 = 12345	1200 = 165432

2. Deferred (Indirect) Addressing

Deferred mode provides an additional level of indirectness. Rather than refer directly to a location to be operated upon, we refer to the address of

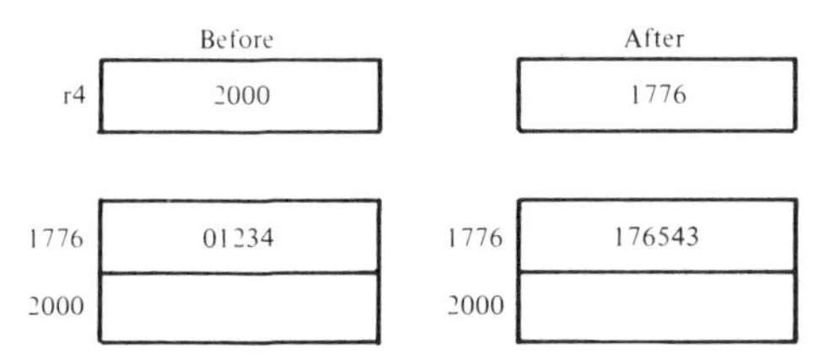

Fig. 2.16. Example of com −(r4).

a location. In the deferred mode, the indirect bit (#4) is set = 1 to indicate deferred addressing. The mode bits are the same as described in direct addressing.

Examples:

(a) Register deferred (mode 001) Symbol: (r) or (R)
inc (r3) 05213

Before	After
1000 = 05213	1000 = 05213
r3 = 1010	r3 = 1010
1010 0000	1010 0001

Note the difference between direct and indirect. In direct mode, "inc r3" would result in r3 becoming 1011 rather than using the contents of r3 as an address to point to the value to be incremented.

(b) Autoincrement (mode=011) and decrement (mode=101) deferred. Symbols: *(r)+ and *−(r); @(R)+ and @−(R).

Table 2.11 shows a comparison of the use of autoincrement for both direct and deferred modes. Autodecrement is similar.

In both the autoincrement and autodecrement deferred mode, the increment or decrement is always increased or decreased by two because the additional level of indirectness references an address and not an operand as in the direct mode.

(c) Indexed deferred (mode=111). Symbol: *x(r) or @X(R).

The concept is the same as for indexed mode except another level of indirectness is added. For example, see Table 2.12.

TABLE 2.11

A Comparison of the Use of Autoincrement for Direct and Deferred Modes

Direct			Indirect (deferred)		
com (r2)+		05122	com *(r2)+		05132
Before:	r2 = 1000		Before:	r2 = 1000	
	1000 2000			1000 2000	
	2000 3000			2000 3000	
Use r2 as address and complement what is in 1000			Use r2 as *address of address* and complement what is in 2000		
After:	r2 = 1002		After:	r2 = 1002	
	1000 175777			1000 2000	
	2000 3000			2000 174777	

TABLE 2.12

EXAMPLE OF INDEXED DEFERRED ADDRESSING

Symbolic code			Machine code		
com *x(r5)			05175		
Before:	x = 200		After:	x = 200	
	r5 = 1000			r5 = 1000	
Address	contents		Address	contents	
1200	2000		1200	2000	
2000	0000		2000	177777[a]	

[a] Complement of 0's.

2.6.2 Double Operand

The format of double-operand instructions is

	Source (SS)		Destination (DD)	
Op code	Mode	R_N	Mode	R_N
15 14 13 12	11 10 9	8 7 6	5 4 3	2 1 0

where the fields are used to specify a source (SS; =from) and a destination (DD; =to). Perhaps the most graphic example is the 'mov' instruction, where one value can be moved from one place to another. For example, mov x,y moves what is in x to y.

Frequently used double-operand instructions are

mnemonic	code	operation
mov(b)	1 ss dd	dd <- ss
cmp(b)	2 ss dd	ss-dd
bit(b)	3 ss dd	ss&dd
bic(b)	4 ss dd	dd <- (!ss)&dd (clears every bit set in ss)
bis(b)	5 ss dd (or)	dd <- ss ‖ dd (sets each bit in dd that is set in ss)
add	6 ss dd	dd <- dd + ss
sub	16 ss dd	dd <- dd − ss

& = and, ‖ = or, <- = becomes, ! = not

A one (1) may be inserted in the numerical code before some of these codes [those with (b) after symbolic mnemonic] to create a byte instruction (e.g., cmpb= 12SSDD and bicb = 14SSDD). The symbol dd ← dd + ss means the destination address receives the value in the destination plus the value in the source address.

Example: add 40(r4), (r1):

Before execution: r4 = 1000
r1 = 2000

Address	Contents
1040	50
2000	20

After execution: r4 = 1000
r1 = 2000

Address	Contents
1040	50
2000	70(sum of 50+ 20)

2.6.3 Use of the PC as a General Register

A common practice is to allow instructions to reference the program counter (PC)—that is, where we are now in the program. Four modes are provided for PC addressing:

010	Immediate	$n
011	Absolute	*$A
110	Relative	A
111	Relative deferred	*A

Immediate mode is a common operation. For example, in "mov $10,r3," the value 10 is moved into r3. The decoding of the statement is

```
mov $10,r3 001  010  111  000  011
             1    2    7    0    3
```

This instruction will actually occupy *two*-word locations, e.g., if loaded at 1000:

```
1000    12703
1002    00010
```

The mov instruction is stored in the first word of the two-word instruction. The second word contains the value 10, the number used in the source specification "$10."

A statement such as "mov x,y" uses the relative mode and will require three words in memory. Imagine that x is located at 1000 and y at 1012 and the instruction "mov x,y" at 1024. To create a test situation, the code listed was typed into a file and assembled i.e.,

```
% make test
I  x:  50
   .=. +10
   y:  0
   .=. +10
   mov x,y
$ex$$
% as test
```

Next, the octal output was examined using the od (octal dump) utility of the UNIX system to generate a listing of the file. A structured permutation of the addresses and instruction codes are shown side by side in Table 2.13

TABLE 2.13

CODE GENERATED FOR PC RELATIVE INSTRUCTION "mov x,y"

```
Source code:   x:      50
               .=.  +10
               y:       0
               .=.  +10
Addresses and instruction decoding:
                    Contents          Symbolic
Address             of memory         code
   0                   50               x:
   2                    0
   4                    0
   6                    0
  10                    0
  12                    0               y:
  14                    0
  16                    0
  20                    0
  22                    0
  24               16767    mov x,y
  26               177750 (-30)
  30               177760 (-20)
  32
Octal dump:
Location
0000000  000407  000032  000000  000000  000030  000000  000000  000000
0000020  000050  000000  000000  000000  000000  000000  000000  000000
0000040  000000  000000  016767  177750  177760  000000  000000  000000
0000060  000000
```

with the instructions. The value 177750 (−30) is stored at location 26, indicating that there is an offset of 30_8 from the PC location (the next address following the value) for the number to be accessed (x). The value 177760(= −20) is placed in location 30 with the PC pointing to 32. Thus, the value −20 is the correct offset to access location 12 ($32_8 - 20_8 = 12$). Note that the values for the mov x,y code would be the same anywhere in memory. The code is position independent since the operands refer only to where the PC is at the current time. "od" (octal dump) produces a header in the first 0–17 locations in the file. Thus, the first octal value represented in the test program is x = 50 at location 20 shown at bottom of Table 2.13.

Although the PC relative instructions are convenient and simple, they are also the slowest instructions. Table 2.14 shows a comparison of the speeds of several example instructions for the PDP-11/03 and PDP-11/34. Examine the tables in DEC-supplied processor handbooks for the execution times of other instructions. Note that register operations are the fastest and that additional levels of indirectness require more processor time.

The total execution time for an "add" instruction for PC relative addressing (mode 7) would be 4.2 + 6.3 + 6.65 μsec = 17.15 μsec for the 11/03 but only = 10.12 μsec for the 11/34. Thus, the 11/03 is about 60% as fast as the 11/34. The newer 11/23 is about 80% as fast as the 11/34.

TABLE 2.14

A COMPARISON OF EXECUTION SPEEDS OF EXAMPLE INSTRUCTIONS ON THE PDP-11/03 AND PDP-11/34[a]

Instruction	PDP-11/03		PDP-11/34A	
Basic times:				
MOV	2.45 μsec		1.83 μsec	
ROR	5.95 μsec		2.18 μsec	
ADD	4.20 μsec		2.03 μsec	
	Source and destination times (word)			
Mode	Source	DST	Source	DST (core memory)
0	0	0	0	0
1,2	1.4 μsec	2.1 μsec	1.13, 1.33 μsec	1.62, 1.77 μsec
3	3.15 μsec	4.2 μsec	2.37 μsec	2.90 μsec
4	2.1 μsec	2.8 μsec	1.28 μsec	1.77 μsec
5,6	4.2 μsec	4.9 μsec	2.57 μsec	3.0, 3.1 μsec
7	6.3 μsec	6.65 μsec	3.80 μsec	4.29 μsec

[a] Total execution time = basic time + source time + destination time.

2.7 ODT MICROCODE

On the 11/03, use of ODT microcode (see Table 2.15) will allow the user to try out the machine instructions discussed above. For example,

```
@1000/xxxx 05003          [clr R3]
@1002/xxxx 05203 LF       [inc R3]
@1004/xxxx 0     CR       [halt]
@1000G                    [run program]
@1006                     [it halts]
@R3/00001                 [look at R3]
```

What has been done is to put a three-instruction program in memory at location 1000. Each instruction occupies 2 bytes, hence the address increments by two.

To place a number into memory, the location is typed following the @, followed by a /, which opens the location and displays its contents (indicated by xs in the above example). The @ sign is the prompt supplied by the ODT microcode resident in the 11/03. For example,

@1000/0000

means location 1000 is examined and zero is in this location. At this point you can put a value into 1000, e.g.,

@1000/0000 05003 CR

Now type

@1000/

and

@1000/5203

will appear.

A CR (carriage return) closes the location and a 'LF' (line feed) does also, but opens the next sequential location automatically, i.e.,

@1000/05003 LF

produces

@1002/0000

To run a program, simply type the address at which execution is to begin and a G. "@1000G" will begin executing whatever is at location 1000.

TABLE 2.15

ODT COMMANDS REPRODUCED FROM THE "LSI-11 PDP 11/03 REFERENCE CARD"[a]

ODT COMMANDS

Format	Octal Code	Description
RETURN	015	Close opened location and accept next command.
LINE FEED	012	Close current location; open next sequential location.
> or]	135	Open previous location.
← or —	137	Take contents of opened location, index by opened location plus 2, and open that location.
@	100	Take contents of opened location as an absolute address and open that location.
r/	057	Open location r.
/	057	Reopen last location.
$n or Rn	044 or 122	Open general register n (0–7) or S (PS register).
r;G or rG	073 107 or 107	Go to location r, initialize the bus, and start program.
nL		Execute bootstrap loader using n as device CSR address.
;P or P	073 120 or 120	Proceed with program execution.
RUBOUT or DELete	177	Erase previous character. Response is a backslash \ (134) each time RUBOUT is entered.
M	115	Maintenance. Display of an internal CPU register follows the M command. Only the last digit displayed is significant, indicating how the CPU entered the Halt (ODT) mode, as follows: Last Digit — Halt Source 0 or 4 — HALT instruction or BHALT L bus signal. 1 or 5 — Bus error occurred white getting device interrupt vector. 2 or 6 — Bus error occurred while doing memory retresh. 3 — Double bus error occurred (stack was non-existent value). 4 — Reserved instruction trap occurred (non-existent Micro-PC address occurred on internal CPU bus). 7 — A combination of 1, 2, and 4 occurred.
CTRL-SHIFT-s	023	For manufacturing tests only. Escape this command function by typing NULL and @ (000 and 100).

After a program is complete (i.e., reaches a halt in our example), registers and memory locations can be examined.

@R6/. . . . .
@R3/00001
@1002/05203

Other functions that PDP-11/03 ODT microcode can perform are shown in Table 2.15.

2.8 CARRY AND OVERFLOW

Carry (C bit in PSW) is defined as the condition that occurs when a statement is executed that produces too many bits to fit in a given word size. Suppose we have a 4-bit computer word and put a $+15_{10} = 1111_2$ in it. If we then add 1, carry occurs:

$$\begin{array}{r} 1111 \\ +\ \ 1 \\ \hline 10000 \end{array}$$

↑
the carry

Overflow (V bit in PSW) occurs when two operands are of the same sign and the result is of an opposite sign.

For simplicity, 4-bit two's complement numbers will be used to explain overflow and carry. Table 2.16 shows all values that are available in a 4-bit word. Table 2.17 shows several simple examples of when the V and C bits

TABLE 2.16

FOUR-BIT TWO'S COMPLEMENT NUMBERS

Decimal	Binary	Decimal	Binary
7	0111	−1	1111
6	0110	−2	1110
5	0101	−3	1101
4	0100	−4	1100
3	0011	−5	1011
2	0010	−6	1010
1	0001	−7	1001
0	0000	−8	1000

TABLE 2.17

EXAMPLES OF CARRY AND OVERFLOW, USING TWO'S COMPLEMENT FOUR-BIT ARITHMETIC[a]

Correct addition			Incorrect addition		
3	011		3	011	
+4	100		+5	101	
7	111	c = 0 v = 0	8	1000	= −8 in two's complement 4-bit arithmetic. Overflow occurs, i.e., v = 1, c = 0.

Other examples

(1)	1	0001	v = 0	(2)	5	101	v = 1
	+2	0010	c = 0		+6	110	c = 0
	3	0011	correct		11	1011(−5)	incorrect
(3)	−6	1010	v = 0	(4)	−6	1010	v = 1
	+ 7	0111	c = 1		−6	1010	c = 1
	+1	0001	correct		−12	0100(4)	incorrect

[a] See Table 2.16.

are set. The carry bit may be used to indicate when the 4-bit word size is of insufficient size for addition. Using an additional 4-bit word will expand the total number of usable bits to 8. Consider adding $49_{10} + 50_{10}$ employing two 4-bit words:

49	0011	0001
+50	+0011	0010
	0110	0011

In this example, neither C nor V are set. However, if

0011	1100	60_{10}
0100	0110	$+70_{10}$
1000	0010	

are added, both the C and V bits are set. The C bit is set, carrying from the first 4-bit word to the second:

```
       1100
      +0110
  1    0010
  ↑
carry
```

Then, for the next word, the C bit is summed

```
 0011
 0100
    1
 ----
 1000
```

The value produced is a negative number, however, since the most significant bit is set. Adding two positive numbers has produced a negative number and the V bit is set to indicate overflow.

The two instructions "adc" (add with carry) and "sbc" (subtract with carry) use the carry bit as well as the operands. Consequently, these instructions are suitable for double precision addition and subtraction. For the PDP-11, it comes as no surprise that a double precision word is 32 bits long.

2.9 GETTING AROUND IN MEMORY

A program usually has a number of parts in addition to the main code, i.e., subroutines, functions, etc. The user normally wishes to move from one part of the code to another. To move from one point to another, branch and jump statements are provided. The mnemonic 'jmp' allows the user to go anywhere in memory, while the branch instructions provide a restricted branch (jump) to ±128 locations from the current location. A simple branch ('br') is allowed as well as many conditional branches (that is, branch if zero, plus, etc.).

Table 2.18 displays the allowable branch instructions ('as' assembler notation). "bes" (branch on error set) and "bec" (branch on error clear) are the same mnemonics as blo and bcs, bhis and bcc. These symbols are available in the 'as' assembler (see Chapter 4) for use in testing error bits returned by system calls in the UNIX system. Note also that certain conveniences are provided in the 'as' assembler such that "unusual" mnemonics such as jle (jump less than equal) can be used in 'as'. 'as' will always generate the code for branch if the target address is close enough. However, if the destination is greater than ±128 locations away, it will generate a jmp. For example, consider the two segments of code

```
         tstb   number
         jle    loc
         .=.+1000
    loc: 0
 number: 0
```

```
         tstb   number
         jle    loc
         .=.+20
    loc: 0
 number: 0
```

TABLE 2.18

BRANCH INSTRUCTIONS[a]

Mnemonics	Octal code	Name Condition
br	000400	branch (unconditional)
bne	001000	br if not equal to zero
beq	001400	br if equal to zero
bge	002000	br if greater or = to zero
blt	002400	br if less than zero
bgt	003000	br if greater than zero
ble	003400	br if less or equal to zero
bpl	100000	br if plus
bmi	100400	br if minus
bhi	101000	br if higher
blos	101400	br if lower or same
bvc	102000	br if overflow (V) clear
bvs	102400	br if overflow (V) set
bhis	103000	br if higher or same
bec (=bcc)	103000	br if carry clear
bcc	103000	br if carry clear
blo	103400	br if lower
bcs	103400	br if lower
bes (=bcs)	103400	br if carry set

[a] bec and bes are included in 'as' assembler to test error bits returned by system calls.

The only difference in these two fragments of code is in the distance of 'loc' and 'number' from the first two instructions. Shown below are the assembled instructions:

```
0   tstb  number      0 tstb  030
4   bgt   012         4 ble   026
6   jmp   *$loc              .
12        .                  .
          .           026 loc:
          .           030 number:
```

In the first case a 'jmp' indirectly to 'loc' is taken if the condition is fulfilled. In the second case, a branch is taken directly to the location since it is within range.

2.9.1 Branch

The simplest branching instruction is br (0004xx) where the operand xx is a number between −128 and plus 127. Branch allows only branching in this range. A branch to a location further away requires use of the jmp instruction. The form of the instruction is

15 — 8	7 — 0
Op code	Word offset

The hardware in the PDP-11 multiplies the word offset by 2 to obtain the number of bytes, and extends the sign into the upper 8 bits of the word. This word is then added to the PC (R7).

Conditional branching is available as shown in Table 2.18. The reader is encouraged to read each branch instruction carefully and to compare this information with the Macro rendition in Table 2.9. A common use of a conditional branch is, for example,

```
loop:   tstb   devstat
        bpl    loop
         .
         .
         .
devstat:
```

Assume "devstat" is the name of a location (16 bit) and that when bit 7 is set the user wishes to proceed (i.e., the user waits until bit 7 is set—"the flag comes up"). Thus, tstb tests the byte shown. When bit 7 is set, the value of the byte is negative:

	15	7	0
devstat:		1	

and a bpl no longer occurs because the last operation produced a negative value.

Branch instructions operate on both signed and unsigned numbers. The instructions

beq	branch if equal to zero	Z = 1
bne	branch if not equal to zero	Z = 0
bmi	branch if minus	N = 1
bpl	branch if plus	N = 0

all operate on signed numbers using single condition codes. The instructions bhi, blos, bhis, and blo operate on unsigned numbers ($0\text{–}177777_8$). The signed conditional branches bge, blt, bgt, ble, operate as follows:

"bge":

PC <− PC + 2* offset if N^V = 0
(^ = exclusive or)

(Recall V is the overflow bit. See section on carry and overflow.) "bge" causes a branch if N or V are both clear or set. "bge" will always cause a branch after two positive numbers are added; i.e., if N = 1 and V = 1, two positive numbers were added and overflow occurred. 'blt' is opposite of 'bge.' That is, pc<−pc + 2* offset if N^V = 1. 'blt' causes a branch following the addition of two negative numbers, even if overflow occurred. 'bgt' and 'ble' operate similarly:

bgt: pc<−pc + 2*offset if Z||(N^V) = 0
ble: pc<−pc + 2*offset if Z||(N^V) = 1
(|| = or)

Consider the following example:

Loc	Label	Symbolic code	Machine code
1000 ------	start:	com r3	05103
1002		.	
1004		.	
1006		.	
1010 ------------------		br start	

What is the code for "br start"? "br start" causes the program to branch back 12_8 bytes (= 5 words):

$$
\begin{array}{rr}
-5_8 \text{ words} = & 177777 \\
& -\quad 5 \\
\hline
& 177772 \\
& +\quad 1 \\
\hline
& 177773
\end{array}
$$

Truncate this word to 8 bits and use bytes: 373 = offset in words, i.e.,

$$
\begin{array}{cccccccc}
 & & & & & \left[\begin{array}{c}3\\11\end{array}\right. & \begin{array}{c}7\\111\end{array} & \left.\begin{array}{c}3\\011\end{array}\right] \\
1 & 111 & 111 & 1 & & & & \\
1 & 7 & 7 & & 7 & & 7 & 3
\end{array}
$$

The code for br is 04xx; packed into a word the code appears as shown in Fig. 2.17.

2.9.2 Jump

The 'jmp' instruction (001DD) allows jumping anywhere in memory. 'DD' specifies the destination address. Suppose one wishes to jump from

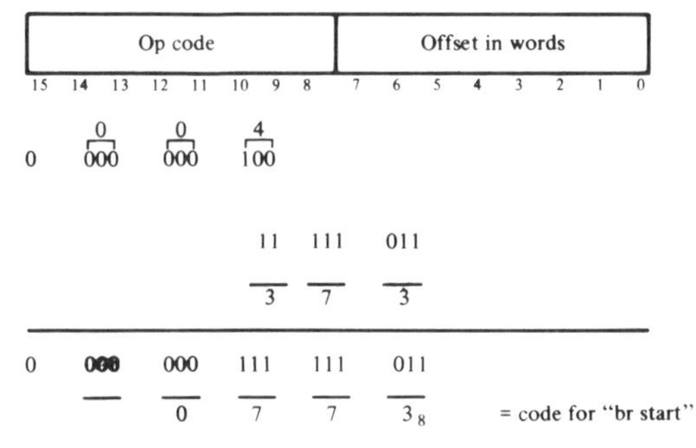

Fig. 2.17. Coding of "br start".

one point to another point in memory:

```
                    .
                    .
                    .
              (one routine)
                    .
                    .
                    .
              jmp another
                    .
                    .
                    .
                    .
                    .
another:            .
                    .
```

The PC is reset to the address of "another" after execution of "jmp another."

Consider the following table look-up example. Depending on the value in r3, the routine shown below will go to the address in "table" corresponding to the value in r3.

```
mov r3, r1          ; r3 contains # (1 to n) of the
                      selected routine.
dec r1              ; range of r3 begins with 1,
                      not 0, therefore
                    ; decrement value in r1
add r1, r1          ; double for bytes
jmp *table(r1)      ; address of table + number of bytes.
```

```
table:  (address of routine 1)
        (address of routine 2)
        (address of routine 3)
                  .
                  .
                  .
```

After "table:" insert the names of routines to be accessed. The assembler will substitute addresses in memory since names are the symbolic references for address. The instruction "jmp *table(r1)" simply adds the value in r1 to the address of table, then jumps to the address in the table at the address formed by the addition.

2.10 SUBROUTINES

The mnemonics

(1) jsr (operand) 04RDD—jump to subroutine, and
(2) rts 00020R—return from subroutine

are used to jump to a subroutine and return from a subroutine, respectively. As shown above, the 6 least significant bits of the jsr instruction are allocated to a destination, and the next 3 bits refer to a register. The return from subroutine (rts) simply specifies an op code and a register in the least significant 3 bits.

The jsr subroutine call makes use of the stack—a space in memory set aside for temporary storage. The stack area is specified by the program by setting R6 to an address in memory (e.g., 1000 in RT-11, 157776 in MINIUNIX). However, the stack may be anywhere so long as adding values to the stack will not write over the program, operating system, device registers, or go outside of memory, which is shown schematically in Fig. 2.18. As more values are stored, the stack grows *downward* in memory (values are said to be "pushed" onto the stack). As values are taken off the stack ("popped") the sp moves upward in memory. Think of the operation as placing plates on a spring-loaded stack in a cafeteria line. Pushing is equivalent to

"mov source, −(sp)"

and popping is equivalent to

"mov (sp)+, dest."

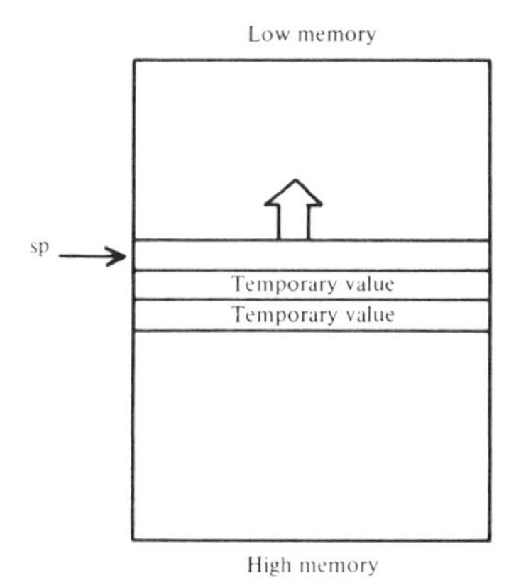

Fig. 2.18. The stack; "sp" is the stack pointer.

The jsr instruction allows referral to a "linkage" register by the convention:

```
jsr r5, subr
   .
   .
   .
   .
subr:
```

When this jsr is executed:

(a) The contents of r5 are pushed onto the stack. This is equivalent to mov r5, −(sp).

(b) The register (r5 here) is loaded with the return address, i.e., the address following the jsr instruction.

(c) A jmp is made to the location named (in this example: subr).

Consider the example shown in Fig. 2.19. After the stack is set to point at main, the value 5 is put in r5 (this is just for illustration). Then, when "jsr

```
main:        mov $.,sp       /Put address of pc in sp, = current pc address
             tst -(sp)       /Decrement sp to allow stack to live
             mov $5,r5       /Below main, put 5 in r5
             jsr r5,sub1     /Jump to sub1
exit1:       sys exit        /Exit
continue:    jsr pc,sub2
             jmp exit1
sub1:        tst (r5)+       /Increment r5 by 2
             rts r5          /Return to continue
sub2:        rts pc
```

Fig. 2.19. Use of the stack to access subroutines.

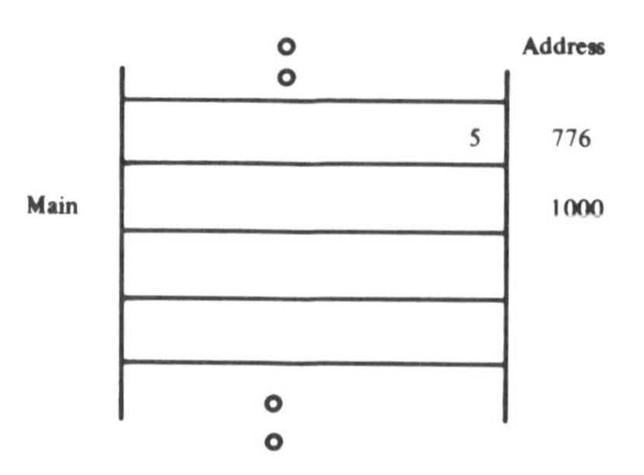

Fig. 2.20. Storage of r5 contents on stack.

r5, sub1'' occurs, the value 5 is put on the stack, e.g., see Fig. 2.20, and r5 is loaded with the return address (here, the address of exit1) and a jump is made to sub1. In sub1, we ''tst (r5)+'' thereby adding 2 to the value in r5 and execute an rts r5. 'r5' now points to 'continue,' and upon execution the program jumps back to 'continue' and is directed to 'sub2' where nothing happens. An 'rts pc' then returns to the jmp exit1 and to the operating system.

Effectively, the 'rts reg' statement can be thought of as

```
mov reg, pc
mov (sp)+, reg
```

The 'rts' instruction simply puts the return address into the PC (thereby returning) and replaces the value stored on the stack back in the register.

When using subroutines, one often wishes to pass arguments. One can either pass values or pointers to values, pointers to pointers, pointers to pointers to pointers, etc. Usually, the most direct thing to do is pass an address (a pointer) of a list of arguments that contains either actual values or pointers to other lists. For example, suppose you wish to pass two values x and y to a subroutine called ''subr'':

```
                  mov value1, x
                  mov value2, y
                  jsr r5, subr
        x:  0      /locations for storing
        y:  0      /x and y follow subroutine call.
  arrayp:  array   /arrayp points
              .    /to a list labelled
              .    /array
              .
```

```
array:
 subr:              /do something with x
                    /and x,y and array
                    /here
          rts r5    /return
```

When subr is entered, x and y may be accessed by the two instructions:

```
mov (r5)+, dest /r5 contains x address
mov (r5)+, dest /r5 contains y address
```

and the first element in the array by

```
mov (r5)+, r3      /r5 contains arrayp address,
                   /address of array
mov (r3)+, dest    /is moved to r3
```

Subsequent values in the array are indirectly accessed by bumping (incrementing) r3 until the end of the array. Note, r5 is left pointing at location following arrayp, a necessity if one wishes to return after the arguments following the jsr. As a specific example of subroutine linkages, consider how arguments are passed using Fortran running under RT-11.

2.10.1 Subroutine Linkage

Fortran subroutines in RT-11 such as "CALL SUBR(arg1, arg2, . . .)" may be accessed via the assembly statement "jsr pc, SUBR" where R5 contains the address of an argument list. The format of the argument block is shown in Fig. 2.21. A-1 is stored in null arguments. Control is returned to Fortran by "rts pc." Similarly, Fortran calls may access assembly level code that is referred to using this format.

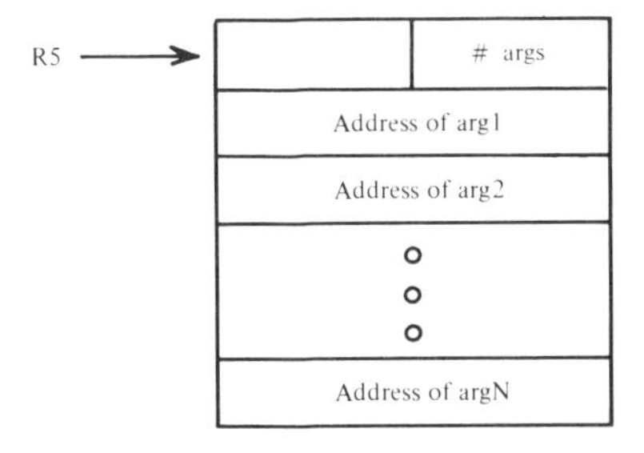

Fig. 2.21. Format of RT-11 Fortran argument block.

Example: Consider the Fortran call

CALL IADD (num, num2, . . . numn, sum)

The assembly routine written in Macro to implement the subroutine IADD is shown below. The example and code displayed are reprinted, with permission, from the Fortran IV Users Guide published by Digital Equipment Corporation. The reader may wish to reread the Macro code below after reading Chapter 4 on "'as' and Macro."

```
        .TITLE ADDER
        .GLOBL IADD
IADD:   MOV (R5)+, R0
        CLR R1
        DECB R0
   1$:  ADD @(R5)+, R1     ;@ is the 'as' equivalent of *
        DECB R0
        BNE 1$             ;1$ is a local label
        MOV R1, @(R5)+
        RTS PC
```


On entry to this routine, the number of arguments is put in R0. Then, the value of the first argument is added to r1 (initially set to zero). Subsequent values are added to r1 until r0 is zero. Finally, the addition is moved to "sum" [MOV R1, @(R5)+] (note that the "@" sign is the Macro equivalent of the 'as' symbol "*"). Two simple examples of the use of this call are shown below.

Example

```
CALL IADD (1, 5, 7, I)
CALL IADD (15, 30, 45, 70, 100, J)
```

In each case all values are added and the result returned in I and J.

2.11 COROUTINES

Coroutines are routines that can pass control back and forth between each other. They are similar to subroutines, except the latter is subservient to a main routine while coroutines maintain a partner relationship. For example, consider the two coroutines shown in Fig. 2.22. One routine

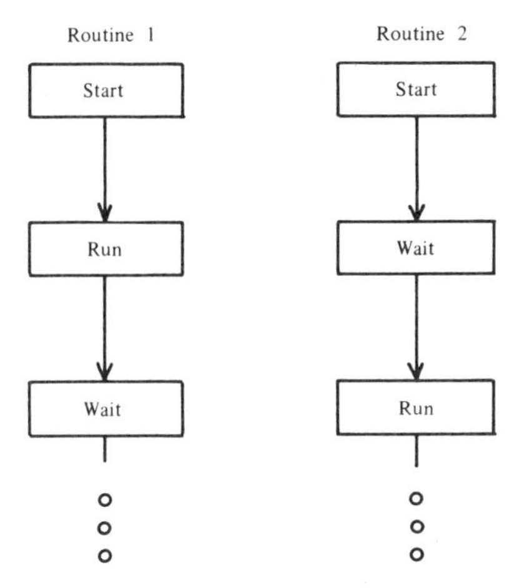

Fig. 2.22. Two coroutines.

runs while the other waits. Control can be passed from one routine to the other by a

jsr pc, *(r6)+

where the entry address of the second coroutine is initially placed on the stack. As the routines pass control to each other, the return values are popped from the stack. For example, when the first

jsr pc, *(r6)+

is executed, the address of the second routine is the destination. Then, when the second routine performs the same statement, the first routine's address is the destination.

Coroutines are useful in programming whenever operations are to be conducted contemporaneously in parallel.

2.12 RECURSION

Another useful feature of stack operation is the ability to easily program recursive operations. Basically, a program may call itself repeatedly placing the return linkage on the stack. On return, the return address is popped off the stack. An example of the use of a recursive routine for writing a short printing routine is given in Chapter 6. The ability to use recursion sometimes results in short and elegant routines.

2.13 EXPERIMENTS WITH INSTRUCTIONS

For each of the PDP-11 instructions given below, determine first what you think the octal code for the instruction should be. Second, using "od" on a UNIX system, check your answer. Third, demonstrate how the instruction works using ODT on an 11/03.

Example:

"add 30(r2), (r1)"

First, you can specify what you think the code for the instruction is from Table 2.9. Second, you can use the assembler ('as') to assemble the code and check whether your coding is correct. Shown below is the sequence of operations necessary to accomplish this compilation. Chapter 4 explains fully what is being done. Third, you can run the program on the 11/03 using ODT as specified in this chapter.

1. code: 066211 for add 30(r2), (r1)

2.

```
% make junk          make a test file using teco
*iadd 30(r2), (r1)   type in the instructions
$ex$$                exit
% as junk            assemble (see Chapter 4)
% od a.out           dump out code
   0. . .
   20 066211 30
```

3. on 03:

```
@R2/----1100      put 1100 in r2
@R1/----1000          1000 in r1
@1000/-----25           25 in 1000
@1130/-----01            1 in 1130
@1010/-066211        66211 in 1010
@1012/-----30           30 in 1012
@1014/------0            0 in 1014 = halt
@1010G            run program at 1010
@1000/0026        (R1) contains answer
```

Determine the octal codes and execute the following instructions:

	Instruction	Conditions
1.	clr r0	let r0 = 10
2.	clr (r0)+	r0 = 1000
3.	add (r1)+, r2	r1 = 1000, r2 = 10, (1000)=20
4.	add −(r1), r2	(776) = 10, r2 = 10, r1 = 1000
5.	add 100(r1), 20(r2)	r2=r1=1000, (1020)=200, (1100)=100

6. dec *(r1)+ r1=1000, (1000) = 2000
7. mov *$10, r0 r0 = 010
8. neg A A = 2000

EXERCISES

1. Convert 1000 (base 8) into decimal.
2. What is the octal value of decimal 4095?
3. If you use an 18-bit computer:
   (a) What are the maximum number of addresses that can be represented with the 18 bits? How does this compare with the PDP-11?
   (b) In two's complement arithmetic, what are the number ranges for integers available with 18 bits?
4. Determine the machine code that should be generated for the following instructions:

   (a) mov #3, r4
   (b) tstb (r4)+
   How do you decide what the machine code for each instruction should be?
5. What does the following code do when executed?

```
012700
000004
000776
```

6. Explain when you would use a jmp instruction instead of a branch instruction. Give an example.
7. In the instruction,

```
tstb x
beq loc
```

   we are testing a specific location in memory. What location is being tested? Which bit in that location must change in order to affect the 'beq' instruction? Why? Would there be any difference if tst was used instead of tstb?
8. When one tests a byte using the "tstb" instruction, where can the negative value be observed?
   Answer: In the processor status word (PSW). Can you prove that this is true by using ODT on the 11/03?

9. Explain the difference between the two statements:

jsr r5, subr

and

jsr pc, subr

What happens in each case when the statement is executed?

10. Suppose you wish to pass 10 values to a subroutine. How can you do it? Give an example.

REFERENCES

A number of excellent texts have been published that relate to the PDP-11. Among these are

Cooper, James, W. (1977). "The Minicomputer in the Laboratory." Wiley, New York.

The use of PDP-11s using Macro in the laboratory is explored. The material is useful parallel reading to this text.

Eckhouse, R. H., and Morris, L. R. (1979). "Minicomputer Systems." Prentice Hall, Englewood Cliffs, New Jersey.

This is a comprehensive text on PDP-11s. Macro is used throughout.

Gill, Arthur (1978). "Machine and Assembly Language Programming of the PDP-11." Prentice Hall, Englewood Cliffs, New Jersey.

This text explores in detail machine and assembly language for the PDP-11. It is highly recommended to readers who wish to expand their knowledge of information given in Chapter 2.

"Introduction to Programming" (1970). Digital Equipment Corporation. PDP-8 Handbook Series, Maynard, Massachusetts.

This handbook contains basic information relating to programming the PDP-8.

MacEwen, G. H. (1980). "Introduction to Computer Systems Using the PDP-11 and Pascal." McGraw-Hill, New York.

This text contains detailed descriptions of PDP-11 fundamentals as well as useful information about programming languages and operating systems. 'C' and the UNIX systems are briefly described.

3

PDP-11 Hardware and Systems

3.1 INTRODUCTION

Describing hardware for the PDP-11 series is difficult since hardware changes are being made at such a rapid rate that it is impossible in any text to give a current description. Rather than try to repeat information that is available in any number of manufacturer-distributed handbooks and other literature, this chapter will discuss some fundamentals, terms, and generally give the reader an overview of PDP-11 hardware and systems.

Digital Equipment Corporation (DEC) has published a number of useful books that describe hardware that is available for both microprocessor systems (11/03, 11/23) and other PDP-11s. The reader is encouraged to examine the handbooks in the microcomputer handbook series. The names of these handbooks change from year to year and contain different amounts of information. One title that has been used is the "Microcomputer Processor Handbook." Other paperback books about peripheral devices are provided to users. These books describe almost anything one would like to know about the 11/03 (LSI-11) computer. The Microcomputer Processor Handbook gives a general description of the LSI-11, including DEC-produced software, modules, the bus, operation, maintenance, and programming. A peripherals handbook is devoted exclusively to a description of all the peripherals and memories that are available for the 11/03. There is also a book published by DEC concerned with peripherals for larger PDP-11s. This publication has been called the "Peripherals Handbook" in recent years. Both books describe the operation of various options that plug into PDP-11s, including how to program each option. However, neither book has many examples and cannot really be employed to learn about how to use the specific devices described. Rather, they should be viewed as reference books.

Each PDP-11 or series of 11s has its own processor handbook. For example, the PDP-11/05/10/35/40 Processor Handbook, PDP 11/60 Processor Handbook, and PDP 11/34 Processor Handbook are various titles in the series. The number of PDP-11s with different model numbers is often bewildering to persons unfamiliar with the 11 series. Thus, a description of these models will be given next.

The PDP-11 series of computers was introduced by Digital Equipment Corporation in the late 1960s and numerous subsequent models have been introduced. All CPUs (central processor units, containing arithmetic elements, etc.) are very similar, and instruction sets are generally compatible across all models. Differences between models are described in the appendices of several of DEC's handbooks. PDP-11 CPU costs now range from several hundred dollars to over $30,000. Needless to say, the capabilities of the more expensive models are significantly better than the smaller models. Table 3.1 shows a listing of the models that have been manufactured in the past decade.

The 11/70 is the top of the line PDP-11 and is often used for time-sharing and real-time applications employing operating systems such as

RSX-11M—a multitasking system (Digital Equipment Corporation),
IAS —industrial applications (DEC),
RSTS —time-sharing; most users use Basic (DEC),
—often seen in business applications,
UNIX* —an excellent operating system for PDP-11s, produced by Bell Laboratories,

RT-11 (real-time 11) is probably DEC's most popular single-user operating system. It is primarily used on small 11s where time-sharing is not employed. PDP-11 11/03s frequently use RT-11 as an operating system. One company (S&H Computer Systems; Nashville, Tennessee) produces a program called TSX, which will allow time-sharing with RT-11. RSX-11M is often the choice for use where multiple tasks are to be contemporaneously accomplished. The UNIX systems can be run on all 11s ranging from 11/03 to 11/70s. The MINIUNIX* system is a small version of the UNIX system that can run on nonmemory-managed PDP-11s. Naturally, the capabilities of the systems are greatly enhanced with larger capacity hardware systems.

A significant problem that the user of DEC equipment faces is the very rapid changes in models and availability of different types of peripherals.

* UNIX and MINIUNIX are trademarks of Bell Laboratories.

TABLE 3.1

PDP-11 MODELS

PDP-11 model number	Comment
05/10	The PDP-11/10 was an early end-user model. The PDP-11/05 was the same computer sold to OEMs (original equipment manufacturers).
15/20	Another early model: 20-End user, 15—OEM. Most early models supported up to 32K words of memory.
35/40	An older version of the 11/34—up to 128Kw.
34	Supports up to 128K words of memory. A popular minicomputer, comparable with a middle of the line Chevrolet in the automobile market.
04	Same as 11/34 but supports only 32Kw.
03	Microcomputer: up to 32Kw.
LSI-11	The microprocessor card in the 11/03. Full-width board (quad) with 4K words of memory.
LSI-11/2	Half-width version with no memory, CPU only.
LSI-11/23 and PDP-11/23	Microcomputer supporting up to 128Kw.
44	A newer version of the 11/34, capable of supporting up to 1M words of memory.
45	128Kw older machine, preceded 11/55.
55	A high-speed processor. Advertized for Fortran processing.
60	Slower than 11/70 but faster than 11/34.
70	The largest PDP-11, often used for time-sharing.

If the characteristics of systems marketed are not understood in detail, the uninitiated user will be unable to choose intelligently a system for use. A common problem is that many potential users are unable to purchase the optimal equipment because they lack the technical knowledge necessary to select a useful set of hardware and software. This book, in part, attempts to guide the user toward proper selection of both hardware and software in laboratory situations.

The fact that the entire PDP-11 series has a common 16-bit architecture is appealing; most instructions are the same for the entire 11 series. This feature is an outstanding characteristic of the PDP-11 series. The machine architecture features a *16-bit word length* (other machines feature 8, 12, 18, 32 bits, etc; generally larger computers use longer words). Different architectural designs are used in various computers. For example, the PDP-8 uses a system in which three buses (multiple information carrying lines) connect the basic elements in the CPU. The control logic, arithmetic logic, memory, and registers are all connected via this set of buses. In contrast, the 11 series uses a single bus, described in the following sec-

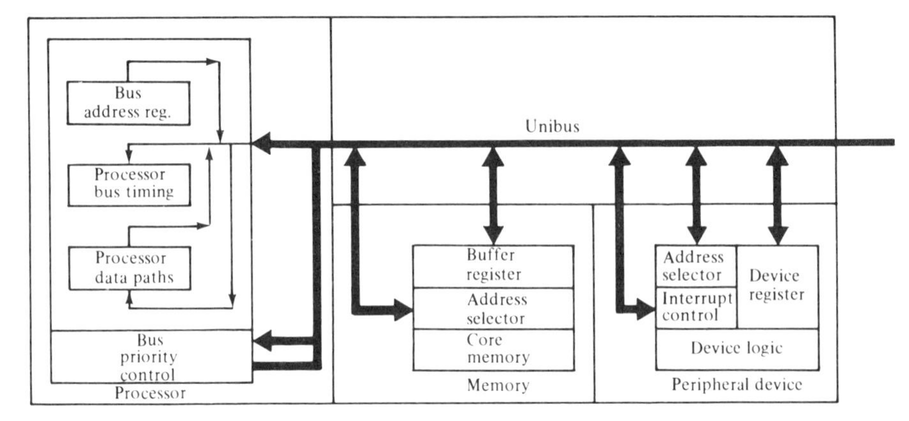

Fig. 3.1. The Unibus. Reprinted from the "PDP-11 Processor Handbook." Copyright 1979, Digital Equipment Corporation, all rights reserved.

tion. Devices all communicate along this bus and can be conceptualized as all "hanging" on the bus as shown in Fig. 3.1. A detailed discussion of the LSI-11 bus is contained in the "Microcomputer Processor Handbook."

3.2 THE PDP-11 BUS

The CPU, memory, and all peripheral devices communicate with each other along a common bus. On the PDP-11/03, this bus contains 16 data/address lines and 17 control and synchronization lines. On larger 11s there are 51 bidirectional lines and 5 unidirectional lines. The bus with fewer lines is frequently referred to as the "Q bus" and the bus with 56 lines as the "Unibus." Devices maintain a master–slave relationship; that is, one device has complete control at any given time. There are two situations in which devices can request control of the bus:

1. Nonprocessor requests (NPR), in which devices transfer data independent of the CPU.
2. Bus requests (BR), in which a device requires servicing. For example, a device may require time-critical servicing and request immediate use of the bus. An example would be when a clock signals that the time has come to take a data sample, trip a relay, etc.

Figure 3.2 shows the priority structure for the 11 series. The processor contains a bus arbitrator. When any device requests service, its priority is determined by where it is located in Fig. 3.2. Five physical priority levels

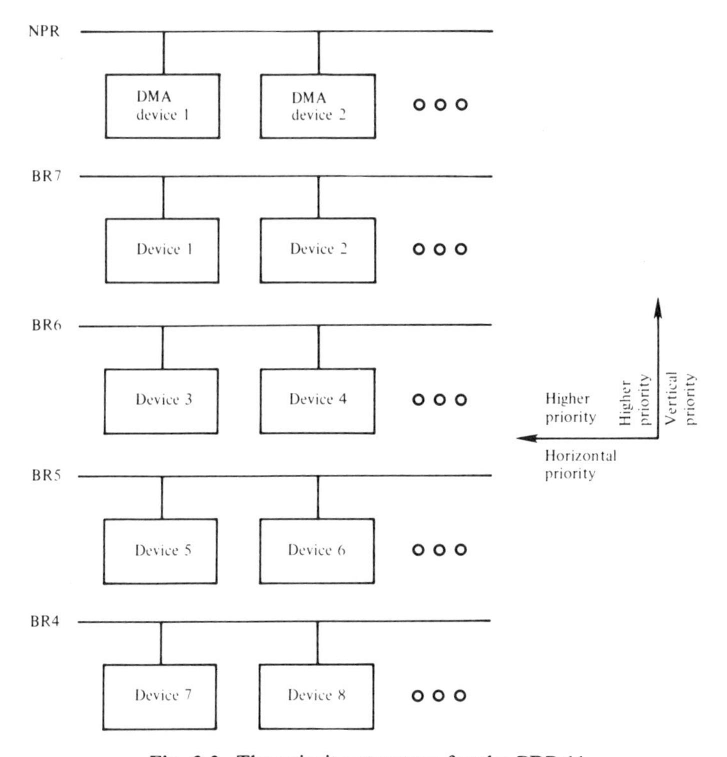

Fig. 3.2. The priority structure for the PDP-11.

can be used. The NPR requests are at the highest level; slow devices, nontime critical devices such as terminal-like devices are at the lowest level (BR4: bus request, level 4). Since only five levels of vertical priority are available, several devices may be assigned to the same level on each horizontal line. The device physically nearest the CPU has the highest priority. Levels BR3–BR0 are unused by peripheral hardware but can be employed by software routines. On 11/03s, the various vertical levels for bus segments are not available. Instead, all devices compete for servicing based on physical distance from the CPU.

As indicated previously, the three bits in the processor status word can be set to control CPU priority. Thus, the CPU can be programmably set to any of eight levels. This feature allows the CPU priority to be changed as desired; for example, to ignore bus requests at a certain level.

After a device requests control of the bus and receives control, it becomes the bus master. When the current bus master completes its activity, it relinquishes control by clearing a bus busy line. In the case of bus requests, a program counter service routine location and processor status word value are loaded into the PC and PSW. For NPRs, data transfers begin between the master and slave. After the NPR transfers are set up, CPU execution of instructions can continue contemporaneously as data are transferred.

3.3 BACKPLANE SIZE AND POWER SUPPLY CAPABILITIES

Computer users often wish to attach more peripherals to a system than there is available physical room. Therefore, the size of the backplane should be considered when a computer system is purchased. It is best to determine initially that there is adequate space and power. If there is insufficient space or power, the manufacturer will be happy for you to purchase additional support hardware.

Computers are generally configured such that cards can be plugged into a slot that is attached to the computer bus. Cards that plug into PDP-11s are referred to as hex, quad, dual- or single-height cards. Figure 3.3 is a line drawing of the outlines of these cards. Figures 3.4 and 3.5 are photographs of actual cards used in an 11/03 system. Figure 3.4 shows a full

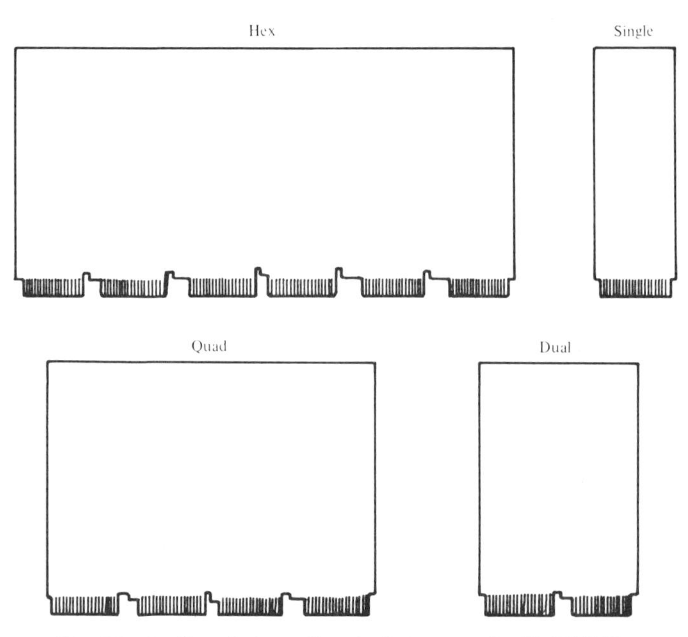

Fig. 3.3. Outline of sizes of cards that are used with PDP-11s.

Fig. 3.4. Photograph of a quad-width LSI-11 CPU card with 4K words of memory.

width (quad) LSI-11 CPU card with 4K words of memory and Fig. 3.5 shows a variety of cards manufactured by MDB systems for use with LSI-11s. Larger PDP-11s will accept any width card (up to hex), but LSI-11s accept only quad cards while LSI-11/2s can use only dual height cards. Further, quad cards that plug into PDP-11 backplanes cannot be directly plugged into LSI-11 quad slots. However, some manufacturers have capitalized on this noncapability by producing cards that will allow Unibus cards (i.e., quad cards for larger 11s) to be plugged via a ''Univerter'' into the Q bus of the 11/03. Able Computer Technology is one such company. Able provides a variety of cards for the 11 besides the Univerter, including a cache memory card that will produce faster execution of programs. Able also produces a card to multiplex several serial input lines. Another company, MDB Systems, produces a series of cards that will connect peripherals such as line printers, communication interfaces, etc., to PDP-11s.

To understand physical limitations of the backplane, consider a standard LSI-11 configuration with quad slots. This system appears (without cards) as shown schematically in Fig. 3.6. Figure 3.7 shows a photograph of an actual four-slot system (immediately below the RX01 floppy disks). Also shown just below the first card cage is an expander box that adds an

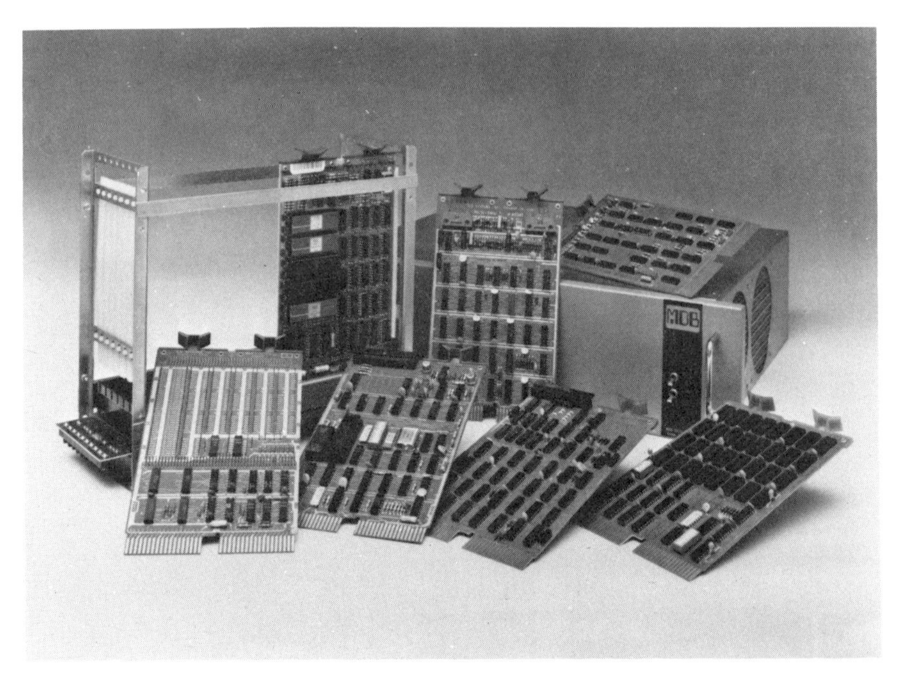

Fig. 3.5. Photograph of interface cards for the LSI-11. Reproduced with permission of MDB Systems.

additional four slots. In this photograph a panel has been added in the rack for parallel I/O, A/D, and D/A connections. The user has four quad slots to use in a minimal PDP-11/03 system. Either one full slot or $\frac{1}{2}$ slot will be used by the CPU ($\frac{1}{2}$ slot for LSI-11/2 or LSI-11/23; 1 slot for LSI-11), $\frac{1}{2}$ slot

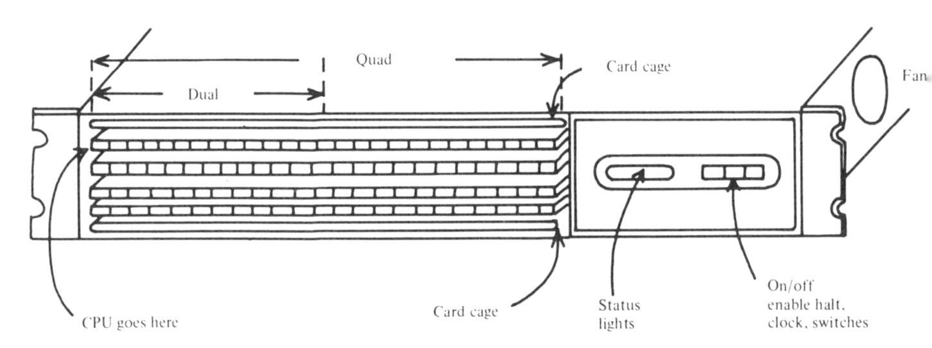

Fig. 3.6. Illustration of LSI-11 card cage and cabinet.

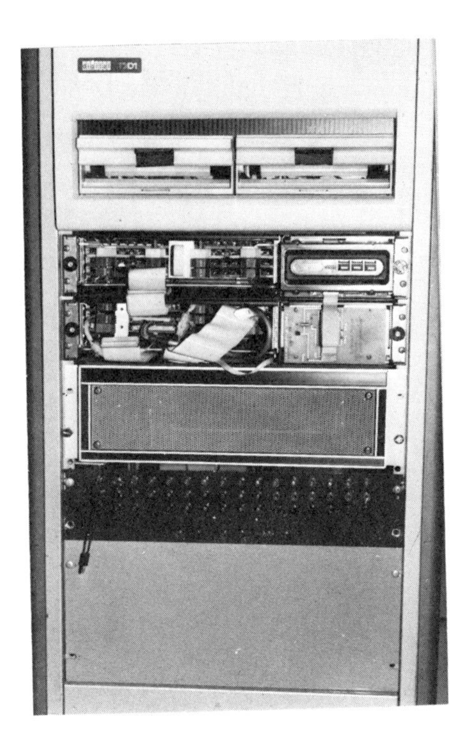

Fig. 3.7. Photograph of a 4-slot LSI-11-based system including RX-01 floppy disks.

for memory, $\frac{1}{2}$ slot for serial I/O, 1 slot for mass storage (e.g., floppy disk), 1 slot for a real-time clock, 1 slot for an A/D and D/A, $\frac{1}{2}$ slot for parallel I/O, etc. The LSI-11 has 4K words memory and requires 1 full slot. The LSI-11/2 requires $\frac{1}{2}$ slot; the remaining $\frac{1}{2}$ slot can be filled with a memory board containing up to 32Kw. Thus, in terms of space the LSI-11/2 is more efficient. For the system elements specified above, there are an insufficient number of slots available in this four-slot system. If we need all the items listed, we must add an *expander box* at additional cost. An expander box contains additional slots and a power supply. Using the DEC handbooks concerned with memories and peripherals, the power supply capabilities must be examined and totaled as devices are added. It is worth noting that some companies (e.g., MDB systems) deliver LSI-11s standard with 8-quad slots. A photograph of an MDB backplane/cardguide assembly is shown in Fig. 3.8.

Although this example is given for an LSI-11, the ideas are similar for larger 11s. For example, backplane units with additional slots (e.g., DD-11C: 2-hex and 2-quad slots, DD-11D: 7-hex and 2-quad slots) are typi-

Fig. 3.8. Eight-quad slot backplane/cardguide assembly for LSI-11 manufactured by MDB systems. Reprinted with permission of MDB Systems.

cally seated on the base of a cabinet such that boards are vertically oriented with their component side parallel to the front of the cabinet. In a pullout drawer in a rack, 20 or more boards (hex or quad) can usually be mounted.

Since adding additional slots and power is expensive, it is prudent to use boards for memory that contain as much storage as possible. While it may be expedient to purchase a single 16K board, it is more efficient to buy a 128K board since both may use the same space in the DD-11.

3.4 BRINGING UP A SYSTEM

In early minicomputer systems, few provisions were provided for interfacing the computer to the user. In such systems the user was required to toggle in a bootstrap program via the switch register on the front panel of the machine. In early PDP-11s, a set of 16 switches (the switch register) was provided on the panel. Raising a switch would set a one in the position equivalent to the physical location of the toggle switch in the row of switches, and lowering the switch would set a zero. After setting an address in memory by toggling in an address (and pressing a switch called, for example, "address load"), instructions could be entered into memory. This procedure proved to be exceptionally cumbersome. Some systems required toggling a short program that would read a tape from a paper-tape reader. Such readers, now obsolete, were standard on early terminals (e.g., ASR 33) used with minicomputers. Paper-tape readers are now rarely produced in combination with terminal devices.

The procedures outlined above for initially getting the computer systems to run are called *bootstrapping*. Systems now use much simpler methods for bootstrapping. Typically, programs for bootstrapping are stored in a read only memory (ROM) and the user presses a button (or toggles a switch) on the front panel that typically reads: "boot." A program in ROM is then executed and usually asks the user from which device information is to be read. On PDP-11s, booting the system will result in a series of diagnostic programs being run and a prompt sign being displayed on the terminal. The user is then required to type in a code for the device on which the operating system is stored. For example, the procedure for bringing up a PDP-11/34 with RP04 disk drives is

(1) press boot on the console,

(2) $DB typed by user—refers to RP04 ($ is a prompt character supplied from ROM),

(3) @unix (Return) ("@" is the prompt supplied by the UNIX system boot),

(4) login: [ready to go].

After the $ sign, the user may specify any device he pleases that is understood by the program running in the ROM. For example, DB refers to an RP04, DK to an RK05, DM to an RK06, etc. In (3), a small loader program (bootstrap) is read into memory and asks the user for the name of the program (in this case, the UNIX operating system) to be loaded. The boot reads the first block (512 bytes) from the disk into memory and executes the code. Thus, each device should have a code at the beginning of the physical device that will allow the remaining information on the disk to be read.

To specify what code is to be inserted in block zero of any given device requires some knowledge of how machine instructions operate the devices and the characteristics of the operating system being used. In the documentation supplied with UNIX system, there is a discussion of how to bring up systems on specific hardware. In the Appendix there are additional comments on how to use several devices that you might happen to have on your system.

For the 11/03, there is a resident ODT (octal debug) that provides the user with certain facilities (see Chapter 2). "resident" means that the program is in ROM. When the power for an 11/03 is turned on, an @ sign appears on the terminal connected to a serial card whose physical jumper connections for the device register are set to locations 177560–177566 for the console device. After an '@' sign appears, memory can be examined, code typed in, device registers examined, etc. In a typical floppy-disk-based system (e.g., an 11/03 with a pair of floppy disk drives), a number of methods exist for booting the floppy. In one, the technique is just the same as on the larger 11s. That is, a prompt ("$") is given and 'DX' is typed to boot the floppies. In another system, a boot is read from block 0 of the floppy and the system booted by starting the program loaded from the floppy. For example,

(1) Press "initial program load" on the floppy disk drive (loads program in memory at 0).
(2) @ 0G (go at 0)

.

.

.

% (The UNIX system prompt appears).

3.5 MEMORY UTILIZATION

Figure 3.9 shows a diagram of how memory is configured on the PDP 11/03. Memory between locations 0 and 400_8 is used for interrupt and trap vectors (see Chapter 8). Between 400 and 160000 programs may be loaded into memory. 160000 to 177776 is reserved for peripherals such as those discussed in Sections 3.6 and 3.7. Depending on the actual amount of memory installed in the system, one may or may not be able to put programs into all locations between 400 and 160000. Memory boards can be purchased with either 4K, 8K, 16K, 32K words. Unless 28K or 32K words are installed, all the locations available will not be physically accessible.

Memory for the PDP-11 is in 16-bit words, but memory addresses are referred to by bytes (2 bytes = 1 word for the 11). Thus, 32K words of memory contains $200{,}000_8$ addresses rather than $100{,}000_8$.

Programs can be written that will run anywhere between 400 and

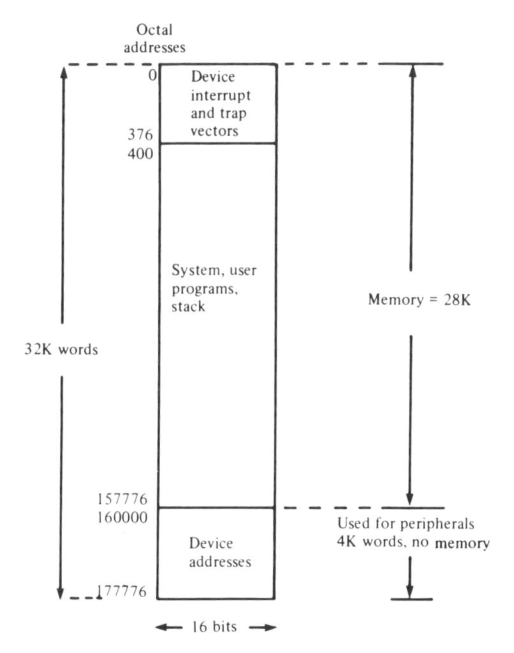

Fig. 3.9. Memory configuration on a PDP-11/03.

157,777 if sufficient memory is plugged into the machine. For many standalone control and small applications, only 4K or 8K words of memory may be necessary. Programs may run stand-alone (that is, with no support from an operating system) or an operating system may reside in memory, as well. Different operating systems provide different facilities to the user and, consequently, use different amounts of memory.

For example, the UNIX system uses about 21K words and MINIUNIX approximately 10K words, while RT-11 requires less than 4K words of memory. Memory costs at the beginning of the 1980s are somewhat less than $1000 for 32K words of memory on a single board. Various manufacturers provide memory boards (e.g., DEC, Motorola, Intel, Mostek, Dataram, etc.) This cost compares to several thousand dollars for 4K words in the early to mid-1970s. If more memory can be easily purchased, it is an inexpensive addition to a system that will allow much more flexibility.

PDP-11s (11/23, 11/34, 11/55, 11/60) with memory management provide the capability of accessing 128K words of memory. Single memory cards for these systems may be purchased containing up to the maximum allowable memory of 128K words. This ability allows use of operating systems with time-sharing capabilities to run in the larger memory space.

3.6 PERIPHERAL DEVICES

Each device (peripheral) that is attached to a PDP-11 bus may require specific hardware modifications to the interface prior to use. All devices must have a specific address set for the device registers and the interrupt vector. Most devices are preset to specified locations (see lists in the Microcomputer Handbook and PDP-11 Peripherals Handbook). For example, the RK05 disk drive is routinely configured to have device registers starting at 777400. Many devices, however, can be set to the address you pick by physically setting jumpers on the card. Examples of such devices are cards for A/D, D/A, clock, serial, and parallel I/O. After all jumpers are set (or left at factory set values), the programmer *must* note what the locations are so that the devices can be accessed from programs. Typically, a name is assigned to a device register location at the beginning of a program. The operating system should contain definitions for most commonly used devices. However, other devices such as those listed above must be added by the user. Some cards will arrive without any jumper settings requiring the user to understand how to select an appropriate address. Such selections can be made by referring to the list of assigned addresses provided by DEC in the handbooks listed above and new device register locations added in unused locations.

What are the various peripherals that can be connected to PDP-11s? The list is rather long—a current listing of some available hardware is shown in Table 3.2. If any device you wish to connect is not available, interface cards are also available to connect nonstandard devices to the 11 bus. For example, DEC provides what it calls a "foundation module" that can be used for custom applications. The module is LSI-11 bus-compatible and contains the logic necessary for communicating with the LSI-11 bus. Seventy-five percent of the quad-height board contains plated-through holes that accept integrated circuits and wire-wrap pins that can be used to construct user circuits. Up to 50 integrated circuits can be contained on the board.

Another approach to custom applications is to use a parallel digital I/O (DI/O) such as the DRV-11C to communicate with nonstandard devices. For example, an A/D module can be interfaced directly to the DI/O board. Alternatively, an A/D chip with ancillary circuitry could be placed on the foundation module. The choice is up to the user. In practice, the

TABLE 3.2

A LIST OF SEVERAL COMMON PERIPHERALS AVAILABLE FOR THE PDP-11 SERIES OF COMPUTERS[a]

Device	Description	Memory address (124–128Kw)
DL11-E, DLV11	Serial terminal cards	777560
PC11/PR11	Printer/punch	777550
LP11	High-speed printer	777514
RK06	14-MB disk	777440
RK11 (RK05)	2.5-MB disk	777400
TC11	DECtape	777340
AD11	Analog-to-digital converter	776760
RP04	80-MB disk	776700
KW11P	Clock	772540
TM11	Tape unit	772500
DR11	Parallel digital I/O	772410

[a] There are numerous others.

use of customized cards is infrequently seen since many manufacturers produce boards for most devices.

3.7 CHARACTERISTICS OF SEVERAL COMMON PERIPHERALS

Listed below are descriptions of several common peripherals and their primary characteristics. No attempt is made to be complete; the reader is referred to a peripherals handbook for more details.

3.7.1 Single Asynchronous Serial Line Interface—DL-11

The DL-11 provides connection of a single serial device (usually a terminal) to a PDP-11. Various models are available that provide options such as the ability to control modems for telephones. The speed of communication is termed the baud rate and can be set from 110 to 9600 baud for both receive and transmit. The console device is accessed via locations

$$\left.\begin{array}{l}777560\\777562\end{array}\right\}\text{receive}$$

$$\left.\begin{array}{l}777564\\777566\end{array}\right\}\text{transmit.}$$

and additional devices for more terminals may be added at locations

776 XX0
776 XX2
776 XX4
776 XX6

where the XX value is set by the user physically on each card that is added.

Chapter 6 contains a description of the use of the DL-11. The DLV-11 is a model available for the LSI-11 computer. A four-channel model is the DLV-11-J.

Serial devices communicate on a character-by-character basis. A second major type of communication is block transfer. Block transfer is used in many mass storage devices such as disk drives and tapes.

3.7.2 Disk Drives

A variety of disk drives is available for PDP-11s, both from DEC and from a number of other manufacturers. Storage sizes range from a minimum of ~2.5 megabytes up to ~300 megabytes. As would be expected, price increases proportionate to the size of the drive. Most disk systems feature a removable cartridge for storage that is divided into blocks and accessed by specifying cylinder, track, and sector. Cylinders can be thought of as vertical concentric sections through stacked platters. Each side of a platter is a track and there are a set number of sectors per track, where a sector is a block (512 bytes). Characteristics of several common disk drives are listed below:

1. *RK05*

Storage size :	2.5 megabytes
Average access time :	50 msec
Data transfer rate :	11.1 microsec/word
Register addresses :	777400–777416

To read and write to the RK05, one must access the register addresses with information including

Number of words to transfer
Address in PDP11 memory
Address on disk
Start/stop/control information
etc.

Normally, device handlers are provided in operating systems to allow access of these registers. Device handlers typically consist of calls to (1) open the device, (2) read from the device, and (3) write on the device. A common configuration for a call would be

read "rk05"	
1000	Block number on disk
10	Number of blocks to read
20,000	Location in memory to store data read.

2. *RP04*

Storage size:	80 megabytes
Average access time:	28 msec
Data transfer rate:	2.5 microsec/word
Registers:	776700–776746

3. *RK06*

Storage size:	14 megabytes
Average access time:	38 msec
Data transfer rate:	4.3 microsec/word

3.7.3 Tape Units

In contrast to fast disk drives, tapes are a slow medium, relatively inexpensive with large storage.

1. *TM11*. The TM11 is a ½-in. digital tape unit with the capacity to store 5 to 20 million characters.

Data transfer rate:	36000 characters/sec
Tape rate:	45 inches/sec (ips)
Registers:	772520–772532.

2. *TU-58* (*Dectape II*). The TU-58 is a very-low-cost cassette tape that uses a microprocessor to control serial communication with a host computer. It is a random-access, fixed-length, blocked-oriented device.

Data transfer rate:	41.7 microsecond/data bit
	= 24 kb/sec
Average access time:	9.3 sec
Tape rate:	30 inches/sec.

From the above it is easy to see the trade-off between tapes and disks in terms of speed and storage capability. Costs on tapes vary from a low of around $500 (TU-58) to around $10,000 for a full size 800 bpi/45ips tape unit such as the TM11.

3.8 THE PDP-11/23

The combination of a dual-height LSI-11/23 processor card, memory card and supporting power supplies, backplane, cabinet, and other accessories is a complete computer system known as the PDP-11/23. While the LSI-11/23 is physically similar to the LSI-11/2, contained on only a single dual-height card, it is functionally equivalent to the PDP-11/34. The most useful feature of the 11/23 is the standard memory management unit (MMU) that allows access of up to 124K words of memory (the upper 4K of memory remains reserved for a peripheral access). In addition, four levels of vectored interrupts are provided in contrast to the on/off interrupt facility of the 11/03 (see Section 3.2 and Chapter 8). This capability allows the 11/23 to be truly useful in a computing environment in which different devices and/or tasks are assigned differing priorities. The 11/23 would be suitable for use with a number of time-sharing operating systems and in multitasking systems in which computing resources with different priorities must be arbitrated. The 11/23 also contains a floating-point emulator (simulates floating-point hardware) chip which is functionally the same as the 11/34 hardware floating point unit. This section describes the main characteristics of the 11/23 including the memory management unit.

The 11/23 can be reasonably used (1) as a laboratory computer in real-time, stand-alone situations, where large amounts of data need to be stored, or (2) in time-sharing or multitasking applications. The availability of large amounts of memory for data storage is frequently appealing in situations such as high-speed data sampling. For example, consider the problem of automated speech recognition in which sampling rates as high as 10 kHz or more may be required. Using a 32K word memory for program and data storage severely limits the number of samples that can be stored. The availability of 96K words for additional data storage provides storage of 192 thousand 8-bit samples or, at a rate of 10K samples/sec, storage of almost 20 sec of information would be possible.

The 11/23 offers a single-chip floating-point unit as an option as do the LSI-11 and the LSI-11/2. It should be understood that not all floating-point units (FPUs) are the same and, in fact, the FPUs on the 11/03 and 11/23 are quite different. The 11/03 floating-point emulator chip simulates the hardware FPU on the PDP-11/40, while the 11/23 unit simulates the 11/34s

FPU characteristics. The 11/03 floating-instruction set (FIS) contains only four instructions (add, subtract, multiply, and divide), while the 11/23 FIS is considerably larger, adding numerous other instructions (e.g., compare). Further, 11/23 instructions can use both single- and double-precision representation (i.e., either two or four 16-bit words represent a floating-point number). The format for floating point is fully explained in literature provided by DEC (e.g., in the "Microcomputer Processor Handbook"). Clearly, the user must be aware of the differences between FPUs when generating code to run on an 11/03 or 11/23 since the same code using floating-point notation will not run on both.

The speed of the FPU on the 11/23 is about 5 to 10 times faster than executing simulated floating-point instructions without the FPU chip. This execution speed, however, is considerably slower than an 11/34 using a hardware FPU.

An appreciation of how execution speeds differ between various floating-point units can be obtained by examining the different execution times required for the "mulf" instruction. Multiply floating (mulf) is an instruction that allows single-precision multiplication (32 bits) of two floating-point numbers. On the FP11-A and FP11-C hardware floating-point units, a double precision (64-bit) "muld" multiply instruction is provided. The FP11-A is a medium speed FPU commonly found on the PDP-11/34 and the FP11-C is higher speed. For register-to-register operations, the FP11-A performs the mulf instruction in 13.4 μsec. The FP11-C is found on larger PDP-11s such as the 11/55 and 11/45 and operates in parallel with the CPU to produce enhanced computation times. The "mulf" instruction typically executes in 2.5 μsec.

The 11/03 and 11/23 have simulated hardware floating-point units and, consequently, have considerably longer execution times. The "mulf" instruction for the 11/23 requires 80 μsec. On the 11/03 an instruction "FMUL" is available, compatible with the 11/40 floating-point unit, and requires ~100 μsec to execute.

Execution speeds for floating-point instructions vary widely, ranging in the above example for a floating multiply from 2.5 μsec to ~100 μsec. Costs for the floating-point units also vary over a similar range from a low of a few hundred dollars for the slower devices to several thousand dollars for the fast hardware.

3.8.1 The 11/23 Memory Management Unit (MMU)

The LSI-11/23 card contains a single-chip memory-management unit that permits access of the entire 128K word memory space. Sixteen bits will allow access of only 64K byte memory locations. PDP-11s without MMUs can, in fact, access only 32K words of memory (e.g., the PDP-11/03). Adding two bits will permit addressing of up to 256K bytes. However, on the 11/23, as on the 11/34 systems, a user is permitted to access only 32K words of what is called "virtual memory." Programs always appear to the user to start at location 0 and to be no longer than 32K words. These programs are said to be in "virtual" memory since they are actually relocated by the MMU to locations in physical memory before the program is run. Consider a time-sharing system in which two users each write programs that utilize similar amounts of memory space. Each user observes that his program extends from 0 to less than 32K. If both wish to run their programs at the same time, it is the task of the operating system to relocate each user's program to different places in memory. Determining where the programs are relocated is the business of the operating system. Neither user knows where his program is being run. He can consider that he has a 0–32K word memory area limitation and never worry about the way it is actually run by the operating system.

The memory management unit has two modes:

1. *Kernel mode.* All instructions can be executed in this mode and all memory can be accessed.
2. *User mode.* Locations in memory that can be accessed are restricted and execution of certain instructions are not permitted (e.g., halt).

The kernel mode can be used for a supervisory program (e.g., an operating system), while the user mode can be used by various users in a time-sharing environment. The user mode is useful in separating users from the operating system in a time-sharing system. A user can be blocked from changing other user's programs or the operating system itself. Clearly, havoc could result if any user had the ability to access other user's programs or the operating system.

The MMU will manage up to 256K bytes of memory addressing with up to 32K words for any individual user (virtual memory). *Two* stack pointers are provided: one for each of the two modes. There are two sets of eight 16-bit word pairs called active page registers (APRs) in which information is stored that is used to form the physical addresses produced when programs are relocated. The user's virtual address space is divided into eight segments or pages, each 4K words in length. The APRs contain the values required to relocate each page to physical memory. Relocation of 4K pages does not need to be contiguous in physical memory. One can conceptualize the operation as follows. Each set of APRs simply contains addresses that are added to the virtual address to form a physical address. However, the address in each APR may differ for each page, thus resulting in spreading the program's code out in various physical address. Such splitting-up of code may be efficient in a multiuser system because of differing sizes of user programs. The operating system provides the intelligence that decides what programs go where. Figure 3.10 shows an example of how a 32K-word-user program might be split in eight pieces in various locations in memory.

When the computer is turned on, it is in kernel mode. In a single-user system the mode probably should not be changed to user particularly if the intent is to be able to access all of memory. Selecting modes is accomplished using the upper four bits of the processor status word (PSW). The PSW appears as shown in Fig. 3.11. If bits 14 and 15 are both 0, kernel mode is selected for the current mode, but if these bits are each 1, user mode is selected. When an interrupt occurs, the PSW is fetched from the interrupt vector location plus two (see Chapter 8 for a complete description) and the mode becomes either user or kernel depending on how bits 14 and 15 are set. Bits 12 and 13 contain the bit settings that indicate the previous mode.

In cases in which a memory-managed system is used for storing large amounts of data in memory, the mode should be kept in kernel mode and previous mode set to user mode. To store data, the APRs can be set such

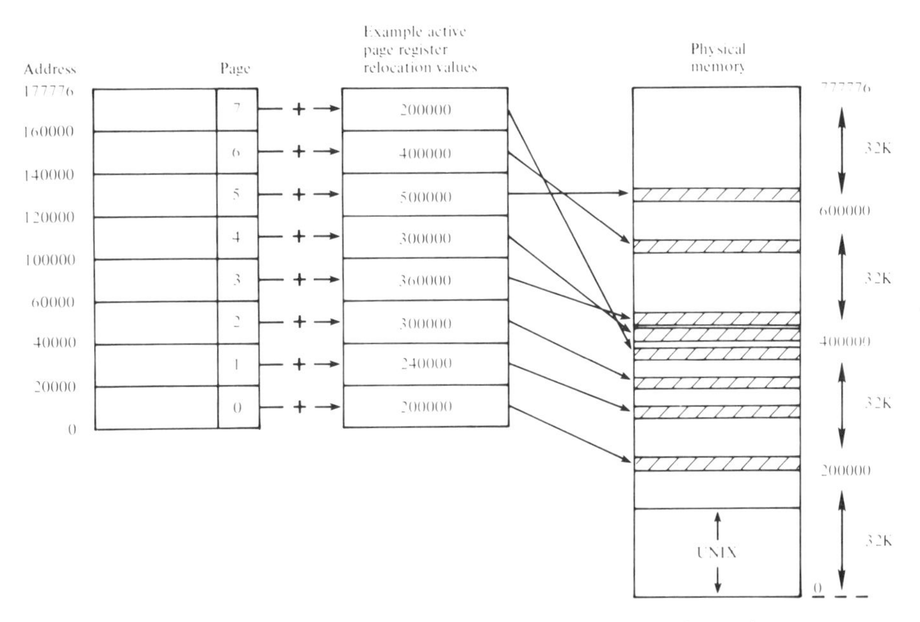

Fig. 3.10. Example of how a user program can be loaded into several areas in memory on a PDP-11/23.

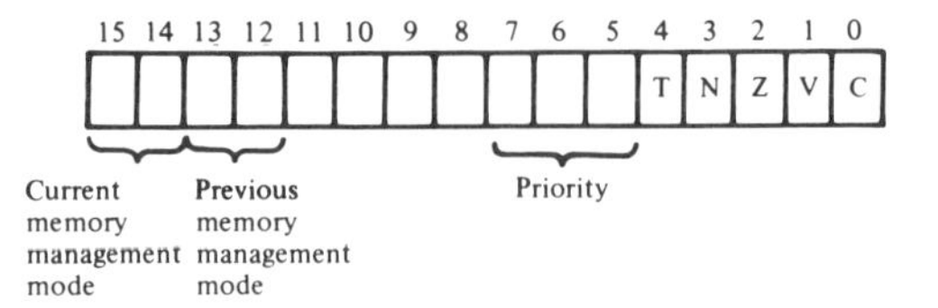

Fig. 3.11. Kernel and user modes in Processor Status Word of memory managed PDP-11s.

that sampled data (e.g., speech data discussed in the example above) is contiguously stored in memory. For example, the eight APRs could initially be set to point to the eight 4K word pages above the program in memory. After these pages are filled, the values in the APRs could be changed to point to the next eight 4K word pages, etc.

The opportunity to change modes comes whenever a software or hardware trap (interrupt) occurs. Thus, when a physical device interrupts (i.e., a clock ticks, etc.) to request servicing, the mode can be set to kernel and the device serviced. Since device locations are in the upper 4K words of memory and not accessible to the users, kernel mode is needed to access device registers. In an operating system, a software trap may occur when a process (running programs, etc.) terminates, or a hardware trap may occur on a timed schedule to allow the operating system the opportunity to determine which program should run next. It is well beyond the scope of this text to discuss operating systems.

From the above descriptions, it is clear that writers of operating systems must understand and deal with the specific hardware characteristics of memory-management units. Of interest to readers of this text is the information that the UNIX operating system will run on the 11/23 in essentially the same way it runs on an 11/34. Thus, the UNIX system can be employed with this inexpensive microcomputer-based system.

Additional information and examples of memory-management usage for the 11/34 are given in the "Microcomputer Processor Handbook" produced by DEC. A discussion of the use of the MMU in an operating system environment is given on a "Commentary on UNIX" written by John Lyons (see Chapter 5 for discussion and reference).

EXERCISES

1. In an application, you find that it is necessary to sample data rapidly from an A/D converter. You have a small computer system and wish to store very long stretches of data that are sampled. With a disk, data can be written rapidly but the small size of your disk precludes its

use. You can, however, purchase a tape drive economically that will store large amounts of data.

(a) For a tape that runs at 45 in./sec, compute the number of 12-bit samples that can be stored per second for a tape with a density of 800 bits/in.
(b) If a tape is

(1) 600 ft or
(2) 2400 ft

long, how much data can be stored for an 800-bpi tape? How does this compare to a disk?

2. Read the description of LSI-11 hardware in DEC's microcomputer processor handbook. Also examine DEC's book concerned with memories and peripherals with special attention to the number and types of peripherals that are available.
3. From the descriptions in (2), specify how to connect a PDP-11/03 to an instrument with an IEEE standard bus. Which card(s) would you need? Are there several alternatives? See if you can compare costs for options you might specify.
4. Obtain a DEC price list. Compare costs for

(a) large and small PDP-11s,
(b) disks, and
(c) tapes.

Plot a graph of dollars (cost) versus number of megabytes stored on several storage devices. To complete this graph, you will need to know size of devices as well as the cost.

5. What is the least expensive way to buy 32K words of memory for a PDP-11/03? What is the most expensive? Is there any advantage to buying more expensive memory?
6. Maintenance contracts are provided by computer companies usually at about 1% of the original cost per month. The alternative is to service computers yourself or pay for service per call. For typical systems you might use, compute the cost of a maintenance contract.

4

'as' and Macro

'as' (for assembler) and Macro are two assemblers commonly used for the generating machine code for the PDP-11. 'as' is a simple assembler that comes with the UNIX* operating system and Macro is an assembler that is included in software packages supplied with most DEC PDP-11-based operating systems. The name "Macro" derives from the assembler's ability to include routines in the assembly code that perform more complex functions than single machine instructions. Such functions are termed "Macros." Macro will run with the UNIX operating system and either Macro or 'as' can be effectively used for assembly of symbolic machine code files.

This chapter will address some of the attributes of 'as' and Macro, especially the differences. The reader is encouraged to read the manuals for these assemblers before proceeding. The 'as' manual is entitled "UNIX Assembler Reference Manual" and was written by Dennis M. Ritchie. This manual is included in the documentation available with the UNIX system. The "RT-11 Reference Manual" and other sources contain information about Macro. Both assemblers are two-pass assemblers. On pass one, all labels are identified, assigned addresses, and symbols stored in the symbol table. On pass two the instructions are assembled using the symbols and addresses assigned in pass one.

Both assemblers are essentially the same except for nomenclature. The differences will be described in this chapter and a number of examples will

* UNIX is a trademark of Bell Laboratories.

be given using simple programs. Since many potential users of the UNIX system are already familiar with Macro, it is felt that a comparison of the two assemblers will provide useful information for persons who have not previously used 'as' and the UNIX system.

4.1 CHARACTERISTICS OF MACRO

1. Statement format. The standard format is

label: operator operand ; comments

Only one statement per line is permitted.

(a) The label must be a user-defined symbol unique within the first six alphanumeric characters. More than one label can be assigned to a line, and each label must be terminated by a colon.
(b) The operator is either a Macro call, an instruction, or an assembler directive. The operand refers to a register either directly or indirectly (modes 0–7).
(c) The comment delimiter is a ';'. All information after the ; is ignored by the assembler.

Example

```
 START: BIT #2, R0    ; BIT TEST
        BEQ LABEL     ; IF PREVIOUS TEST EQUALLED ZERO
                      ; GO TO LABEL
        MOV #10, R3   ;
LABEL:  RTS PC
```

2. The symbols for immediate and indirect (deferred) are '#' and '@', respectively; i.e.,

```
MOV #200, R0
MOV @(R1)+, R0
```

3. Assignment. Symbols may be set equal to expressions; e.g.,

```
A = 10
B = A-2 & MASK.
```

Operations specified in assignment statements are conducted when the program is assembled. Register names are assigned by the convention

R0 = %0
R1 = %1
R2 = %2
.
.
.
SP = %6
PC = %7

4. Operators

(a) For division, the symbol used is '/'. Other standard symbols for addition, subtraction, and multiplication (+, −, *) are also available. Specific mention of the '/' symbol is made because of the differences with 'as' discussed in the next section.

(b) Less frequently used symbols are

Symbol	*Definition*
$	logical AND
!	logical inclusive OR
^	universal unary operator

The user should note carefully that the use of operators is valid only during program assembly and cannot be used for operations in program execution. For example, the '/' symbol cannot be employed to divide two variables while the program is executing.

5. '.' refers to the current location. For example, MOV #., R0 moves the current location to R0.

6. A useful feature of RT-11 Macro is the ability to employ a variety of Macro calls such as

.TITLE	title can be included
.BYTE	bytes follow this directive
.WORD ,5,0	stores 0,5,0 in consecutive words
.ASCII /char. storage/	

For example,

	.ASCII /HELLO/ stores HELLO in consecutive bytes
.ASCIZ /HELLO/	A zero is inserted at the end of the string of bytes stored
.EVEN	Sets PC to the next even address
.ODD	Sets PC to the next odd address
.BLKB	Specifies the number of bytes to be reserved
.BLKW	Specifies the number of words to be reserved

.GLOBL	Define variables globally
.IF	Conditional assembly command
.ENDC	End conditional assembly
.PRINT	Prints ASCII strings
.END START	Specify entry location. This statement is required to be the last statement in a program.
.EXIT	Exits in an RT-11 environment

The above calls are included in a program by simply inserting them at the operator position in the standard format. For example,

```
START:  MOV #10, R0
                .
                .
                .
        .PRINT #LABEL
LABEL:  .ASCIZ /THE END IS NEAR/
        .END START
```

Figure 4.1 shows a simple program at the top of the figure for printing a string of characters using Macro in RT-11. This example shows how Macro calls can be used exclusively to accomplish some operations. A variety of Macro calls are available and are described in the DEC reference manual for Macro.

The ".PRINT" example given in Fig. 4.1 is only relevant to RT-11 Macro unless a Macro call is specifically configured to run under the UNIX system, which will mimic the RT-11 call. Since a Macro is nothing more than a small program or set of definitions, all Macros can be collected in a file and accessed when programs are assembled. For example, in one system running under the UNIX system, a file named sysmac.sml was created in account "/etc". The "param" call in RT-11 Macro was translated as follows:

```
.macro  $param
    r0 = %0
    r1 = %1
    r2 = %2
    r3 = %3
    r4 = %4
    r5 = %5
    sp = %6
    pc = %7
.endm  $param
```

```
;Shown below is an RT-11 example of the use of macros
        .GLOBL HELLO
        .MCALL .PRINT
        .MCALL ..V2..,.REGDEF,.EXIT
        .REGDEF
        ..V2..
HELLO:  .PRINT #LIST
        .EXIT
LIST:   .ASCIZ /HELLO/
        .END HELLO

;Running under Unix a similar program is shown below

        .MCALL $PRINT
        .MCALL $PARAM,$EXIT
        $PARAM
HELLO:  $PRINT HELLO
        $EXIT
        .END HELLO

; The listing file is shown next



1                               .MCALL $PRINT
2                               .MCALL $PARAM,$EXIT
3 000000                        $PARAM
4 000000              HELLO:    $PRINT HELLO
5 000024                        $EXIT
6           000000'             .END HELLO
```

Fig. 4.1. The use of Macros in RT-11 and the UNIX operating system.

and included in this file. Then, if one wished to include the above code in a Macro program running under the UNIX system, the Macro call:

$param

would be inserted at the operator position in the code. For a complete example, compare the RT-11 Macro version of the print program at the top of Fig. 4.1 with the UNIX system version at the middle (code) and bottom (listing) of the figure. The "$PRINT" Macro is written to print the string after the call rather than take the address of the string as in the RT-11 example. The user can create many different types of Macros. At our installation, various students have produced Macros for various purposes.

The Macros used when the UNIX system is the operating system may utilize the system calls in Section II of the Programmer's Manual. For example, to exit from a program and return to the UNIX system, a

"TRAP 1" instruction should be issued. [See exit(II) of the UNIX Programmer's Manual.] To create a Macro for this call, the following code can be included in sysmac.sml:

```
.macro $exit
104400 + 1
.endm $exit
```

TABLE 4.1

SUMMARY OF MACRO SYMBOL RULES

(1) Alphanumeric characters including $ and period in ASCII code are used.
(2) The first character is not a digit (except for local symbols).
(3) The symbols are unique within the first six characters.
(4) There may be no embedded spaces, tabs, or illegal characters.
(5) The symbol value depends on whether it is used as a:
 (a) Macro symbol,
 (b) permanent symbol (mnemonic code), or
 (c) user-defined symbol (internal or global).
(6) The equal (=) sign must be used to separate symbols from expression:
 (a) The assignment operator is found in the operator field.
 (b) Only one symbol may be defined by any one assignment.
 (c) Only one level of forward referencing is allowed.
(7) Register symbols %0 . . . %7. The following symbol assignments are recommended:

```
R0 = %0
R1 = %1
R2 = %2
R3 = %3
R4 = %4
R5 = %5
SP = %6
PC = %7
```

(8) Local symbols may be of the form n$ where n is a decimal integer from 1–127 inclusive.
(9) Period (.) is the symbol for the assembly location counter.
 (a) When used in the operand field, the period represents the address of the first word of the instruction. For example,

```
        . = 500        ; set location counter to 500
A:  MOV #., R0         ; R0 = location A = 500
```

 (b) When used in the operand field of an assembler directive, the period represents the address of the current byte or word.
 (c) In a relocatable program, one must use form . = . + expression. For example,

```
A:   . = . + 100      ; reserves 100 (octal) bytes
                      ; of storage space at the location designated
                      ; by label A
```

(10) All numbers are interpreted as octal radix unless otherwise specified by a decimal point following the number. In this case, the number is treated as decimal. Octal numbers 8 and 9 flag an error and are treated as decimal numbers.

This code corresponds to a trap 1 (see your PDP-11 manual to verify). If the "$exit" Macro is included in your program it will produce an immediate exit from the program and a return to the UNIX system. Note that the code in sysmac.sml is in ASCII.

7. Table 4.1 contains a short summary of symbol rules for Macro that can be used for reference.

4.2 CHARACTERISTICS OF 'as'

1. Statement format. The statement format is the same as Macro except

(a) the name can have eight alphanumeric characters including period, _, $, and ~ ;
(b) digits 0–9 followed by a colon serve as a temporary symbol;
(c) the comment field begins with a "/".

2. Symbols.

(a) Lower and upper case characters are distinguished but only instructions and directives are required to be lower case.
(b) The predefined symbols used for immediate and indirect addressing are "$" and "*", respectively. For example,

```
mov $200, r0
mov *(r1)+, r0
```

(c) Registers are predefined in the assembler as r0, r1, r2, r3, r4, r5, sp, pc.

3. Assignment operations are the same as for Macro.

4. Operators. Since '/' begins a comment in 'as', the symbol '\/' is used to indicate division. '\' is a backslash. Other operators are:

$	Logical And
|	Logical Or
!	Not (different in Macro; same as in 'C')
^	Result has value of first operand and type of second operand, used to define new instructions
>>	Logical shift right
<<	Logical shift left

blank	When there is no operator between operands, the effect is the same as if there was a "+"
+	Addition
−	Subtraction
*	Multiplication
%	Modulo

5. '.' refers to current location, i.e.,

```
mov $., R0
```

ASCII strings may be included by the convention

```
list: <The end is near \n>
```

where the characters within angle brackets are converted to a sequential string of bytes. Characters preceded by a backslash have the following meanings:

\n	new line
\t	tab
\e	EOT
\0	NUL
\r	carriage return

6. 'as' does not utilize Macros per se. However, there are a few useful directives that can be employed in much the same manner as Macros. For example,

.byte expression. . .	expressions are assembled in successive bytes
.even	Sets the PC to the next even location, if odd.
.if "expression"	If "expression" defined, statements between if and endif will be assembled.
.endif	
.globl name. . .	Makes name external (i.e., accessible by other programs).
.text	Causes the assembler to assemble the statements that follow into text (i.e., program statements),
.bss	Data or uninitialized storage (bss).

TABLE 4.2

SUMMARY OF 'as' SYMBOL RULES

(1) Alphanumeric characters including period (.), underscore (_), tilde (~) may be used.
(2) The first character may not be a digit.
(3) The symbols must be unique within eight characters.
(4) Use of tilde (~) in identifier:
 (a) When encountered, it is discarded and the remaining identifier generates a unique entry in symbol table.
 (b) The "C" compiler uses the ~ to place names of local variables in the output symbol table without having to worry about making them unique.
(5) Temporary symbols consist of a digit followed by "f" or "b."

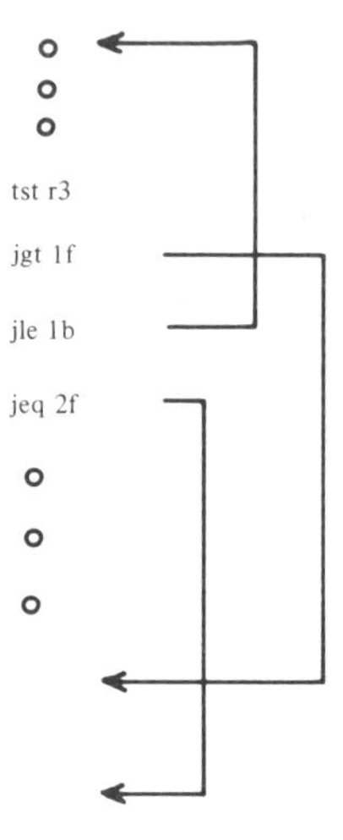

(6) Number constants are interpreted as octal radix unless otherwise specified by a decimal point following the number.
 (a) Octal 8 = 10, and 9 = 11.
 (b) Octal constants are truncated to 16 bits and interpreted in two's complement notation.
 (c) Decimal constants should be representable in 15 bits (i.e., less than 36,768).
(7) Character constants:
 (a) Single character constants are of the form single quote followed by single ASCII character. The value of a constant is its ASCII code value.
 (b) Double character constants are of the form double quote followed by two ASCII characters.
  1. The value of the constant for the first character is in the least-significant byte and the code for the second character is in the most-significant byte.
(8) Register symbols are predefined as: r0, . . . r5, sp, pc. One can also refer to r3 as r0 + 3.
(9) Period (.) serves as a location counter and can be used to reserve locations by resetting its value. For example,

```
A:   . = . + 100     /reserves 100 octal bytes
                      following A.
```

.comm name, expression	Equivalent to .global name: .=.+expression Name is in bss with "expression" bytes reserved.

7. Table 4.2 displays a summary of symbol rules for 'as'.

4.3 'as'/MACRO COMPARISON

The most noticable difference between the symbolic codes for the two assemblers is in the use of symbols for indirect and immediate addressing and the use of lower case letters in 'as'.

The symbols for deferred (indirect) and immediate addressing are shown below:

	'as'	Macro
deferred	*	@
immediate	$	#

Thus the two statements

```
mov $10, *r3      ('as')
mov #10, @R3      (Macro)
```

are equivalent. This convention for 'as' is followed to make 'as' equivalent to the same conventions in 'C'. In 'C', '*' indicates indirectness.

4.4 EXAMPLES

4.4.1 Macro Example

Figure 4.2 shows a simple test program for adding five numbers: 1 + 2 + 3 + 4 + 5 = 15. The program test.mac has been typed in using teco. Note that registers are defined as shown in lines 1 and 2 (unnecessary in 'as'). The program simply puts the address of the list K (#K, line 4) into R1, then adds what R1 points to into R0 (line 5) until (line 6) *10 bytes* (5 words) have been added. After all values have been added, the program returns to the monitor (UNIX) via a TRAP 1 instruction. TRAP 1 is the same as "sys exit" in 'as'. Both activate the same trap and simply provide a path back to the operating system. See Section II of the UNIX manual

```
         R0 = %0
         R1 = %1
MAIN:    MOV #0, R0
         MOV #K, R1
LOOP:    ADD (R1)+,R0
         CMP #K+10,R1
         BPL LOOP
         TRAP 1
K:       1,2,3,4,5
         .END MAIN
```

Fig. 4.2. A Macro program for adding five numbers.

and a description of operating system calls in this chapter and in Chapter 5. After creating the Macro program using teco, i.e.,

```
% make test.mac
                      [type in the program]
$ex$$                 [exit]
```

we next assemble and link the program:

```
% macro -1s test.mac  ["-1s" creates a listing file]
                      [called test.1st,]
                      [test.obj is also created.]
% link test.obj       [creates test.out]
% test.out            [runs program]
```

The listing file that is created is shown in Fig. 4.3. Column 2 shows the addresses relative to 0, and column 3 the code that was generated for that address. For example, addresses

```
0004  012701
0006      22  [refers to location 22]
```

show clearly that the immediate mov instruction refers to location 22 for K.

An object file is created by Macro. Object files are files assembled but not loaded into a specific location in memory. File names ending in .obj conventionally refer to object files. Next, we must link the object file and create a run file (test.out). Linking is used to allow various object files to be concatenated in memory and amalgamated into one file for subsequent running. For example, imagine that you have assembled separately a main program and three subroutines. The first file of each of four routines will appear relative to zero, but when linked they will be concatenated and

```
 1          000000                      R0 = %0
 2          000001                      R1 = %1
 3  000000  012700          MAIN:       MOV #0, R0
            000000
 4  000004  012701                      MOV #K, R1
            000022'
 5  000010  062100          LOOP:       ADD (R1)+,R0
 6  000012  022701                      CMP #K+10,R1
            000000G
 7  000016  100374                      BPL LOOP
 8  000020  104401                      TRAP 1
 9  000022  000001          K:          1,2,3,4,5
            000002
            000003
            000004
            000005
10          000000'                     .END MAIN
```

Fig. 4.3. Listing file corresponding to the code in Fig. 4.2.

located at a particular point in memory (for RT-11, 1000 : MINIUNIX, 60,000). In the UNIX system, the memory management hardware relocates all programs relative to zero allowing any user up to a 32K segment of memory. For example,

% link main.obj subl.obj sub2.obj sub3.obj

creates a file called main.out which contains all the subroutines. "test.out" (equivalent to a.out in 'as') is essentially a "snapshot" of the linked program in memory. Linking produces a file referred to as a "load module" or "run file."

Note that we have a difficulty with our example. When we type

% test.out

nothing happens. The program has no output! How can we tell what is going on internally in the program? There are several ways:

(1) Write an output routine.
(2) Use a debugger to run the program.
(3) Type in octal instructions from .1st file on a PDP-11/03 and run with the ODT microcode.

Methods (1) and (2) will be examined in the next two sections of this chapter. The reader with a PDP-11/03 can try out immediately method (3). An example in 'as' is given next.

4.4.2 'as' Example

The procedure for creating and running an 'as' program is much the same as Macro. Consider the following sequence:

```
% [create file]      [use editor to create prog.s]
% as prog.s          [assemble prog.s]
% a.out              [runs program]
% mv a.out prog      [saves run file in file]
                     [named prog]
% prog               [runs prog]
```

Figure 4.4 shows the same program for adding five numbers as shown previously for Macro. Note the following changes:

1. lower case letters;
2. # becomes $;
3. "TRAP 1" becomes "sys exit";
4. registers are predefined as r0, . . . , r5, sp, pc.

Since we cannot tell if this program runs, additional code needs to be added to allow the answer to be printed out.

Figure 4.5 displays a modified version of Fig. 4.4, adding the facility to print out the value of the summation. The first five lines are the same as before.

1. Line 6 stores the value 5 in "count" (number of characters to be printed out.)

2. Line 7 moves the address to the character buffer +5 into r3. We set up a buffer to hold the ASCII values of the characters. A buffer of five characters provides space the store the five characters that are to be printed out.

```
start:  mov $0,r0
        mov $k,r1
loop:   add(r1)+,r0
        cmp $k+10,r1
        bpl loop
        sys exit
k:      1
        2
        3
        4
        5
```

Fig. 4.4. An 'as' program for adding five numbers. Equivalent to the Macro program in Fig. 4.2.

```
1  start:  mov $0,r0        /r0 = 0
2          mov $k,r1        /address of k in r1
3  loop:   add (r1)+,r0
4          cmp $k+10,r1     /compare the end of the list with r1
5          bpl loop
6          mov $5,count     / five numbers are being added in this example
7          mov $buffer+5,r3         /address of the end of the buffer
8  get:    mov r0,r4        /save answer
9          bic $177770,r4   /clear all the bits except the last three
0          add $60,r4       /convert the number to ascii by adding 60
1          movb r4,-(r3)    / put the value on the output buffer list
2          asr r0;asr r0;asr r0;    /now get the next octal digit
3          dec count
4          bne get
5          mov $1,r0        /1 is standard output, 0 is standard input
6          sys write        /write is entered with the file ID in r0,
7                           /the "file" is the standard output
8          buffer           /the address of buffer is put here on assembly
9          5                /the number of characters to write out
0          sys exit
1  k:      1                / the data list
2          2
3          3
4          4
5          5
6  buffer:
7          .even
8          .=.+20.
9
0
1  count:  0
```

Fig. 4.5. A modified version of Fig. 4.4. The ability to print out the answer is included.

3. Line 8: r0 — the result ($15_{10} = 01111_2$) is moved to r4.
4. Line 9: clear all but last 3 bits.
5. Line 10: add 60 + 7 = 67 to obtain the ASCII code for 7.
6. Line 11: put 67 into the buffer to be printed out.
7. Line 12: shift the next value right 3 bits: i.e., 01111 shifted 3 bits = 01. This is the next character to print.
8. Go back to "get:" and print the character 01.
9. Line 13: keep going until five characters are stored in the buffer.
10. Lines 16, 17, 18, 19: This is the standard write format (look it up in Section II of your UNIX operating system manual). Line 19 = #bytes, line 18 = address of buffer, r0 receives a 1 for writing on the standard output.
11. Line 20: exit.

1. Original value to be printed in octal (r0): 0 0 0 1 7

2. Converted to bytes: 60 60 60 61 67

3. Stored in buffer:

60	60
61	60
	67

4. How the numbers appear as words:

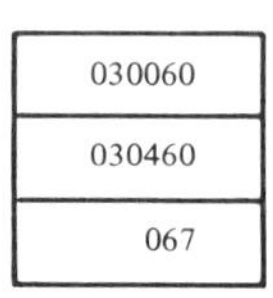

5. Why the words look this way:

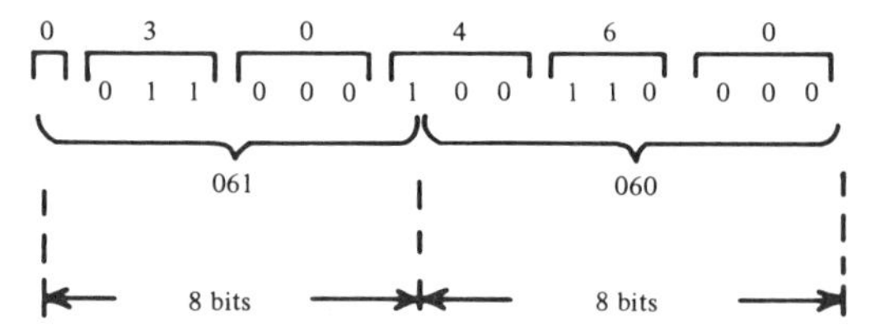

Fig. 4.6. Explanation of data-storage procedures employed in the program shown in Fig. 4.5.

Figure 4.6 indicates how the buffer is filled and printed. A confusing feature in the way words are filled with byte information. The figure shows how bytes 060 and 061 are packed into one word to yield the word value 30460. This packing occurs because we represent octal information in 3-bit chunks. Thus, at the end of the 8th bit we need another bit to fill out the 3rd octal digit.

If we now run test.as after assembly

```
% a.out
00017%
```

is printed.

4.5 TESTING PROGRAMS WITH "ddt" (DYNAMIC DEBUGGING TECHNIQUE)

"ddt" is a system utility program not unlike "odt" in concept. It can be obtained from the UNIX distribution package. "odt" is a system utility used in DEC operating systems such as RT-11. "odt" stands for octal debugging technique and uses only octal addressing and machine code representation. "ddt" differs from "odt" in that symbols can be obtained from the symbol table associated with a file and instructions decoded to print in mnemonic form. It is useful for running programs step by step, setting breakpoints in a program and examining the results of partial execution. "ddt" allows use of symbolic addresses while "odt" uses only octal. We shall next use ddt to run the program in Fig. 4.5.

Table 4.3 shows a summary of some of the more important ddt commands. To examine the contents of your run file, simply type "*% ddt* ," if a.out is the file you are debugging. If another file is to be debugged, type *% ddt filename*. If you have moved the file to test you would, for example, type

% ddt test

Figure 4.7 shows the results of typing

0,045/	[prints the symbolic contents of the file; '/' is the same as /i]
0,057/o	[prints the octal contents of the file]

Compare the symbolic contents with the original program in Fig. 4.5. All the immediate references and relative references are filled with references to locations. Further, the octal code for all locations is available via the /o option. Using this option, you can observe how each instruction is decoded. The values 0 to 045 display the entire program. Note that the number zero must precede the location you are specifying to indicate an octal value. Hence, to print from 0 to 045_8: 0,045/ is typed.

The nm [filename] command will tell you where the labels are located in the program relative to zero. For example, as shown in Fig. 4.8, buffer is located at 110t where t stands for text, d for data, and b stands for uninitialized storage. There is no d or b in this nm listing. Figure 4.9 shows the standard sequence used for assembling the test.as file and running it.

To conduct a debugging of test, first set a breakpoint at location 016 by

```
% ddt test
016; b          and type
0;g
```

```
% ddt test

0,045/i
0          mov      $0,r0
04         mov      $76,r1
010        add      (r1)+,r0
012        cmp      $106,r1
016        bpl      010
020        mov      $5,count
026        mov      $115,r3
032        mov      r0,r4
034        bic      $177770,r4
040        add      $60,r4
044        movb     r4,-(r3)
046        asr      r0
050        asr      r0
052        asr      r0
054        dec      count
060        bne      032
062        mov      $1,r0
066        sys      write
070        jmp      *r0
072        reset
074        sys      exit
076        wait
0100       rti
K+04       oct      03
K+06       iot
K+010      reset
buffer     halt
buffer+02           halt
buffer+04           halt
buffer+06           halt
buffer+010          halt
buffer+012          halt
buffer+014          halt
buffer+016          halt
buffer+020          halt
buffer+022          halt
count      halt
```

Symbolic Interpretation

```
0,057/o
0          012700
02         0
04         012701
06         076
010        062100
012        022701
014        0106
016        0100374
020        012767
022        05
024        0106
026        012703
030        0115
032        010004
034        042704
036        0177770
040        062704
042        060
044        0110443
046        06200
050        06200
052        06200
054        05367
056        054
060        01364
062        012700
064        01
066        0104404
070        0110
072        05
074        0104401
076        01
0100       02
K+04       03
K+06       04
K+010      05
buffer     0
buffer+02             0
buffer+04             0
buffer+06             0
buffer+010            0
buffer+012            0
buffer+014            0
buffer+016            0
buffer+020            0
buffer+022            0
count      0
```

Octal Representation

Fig. 4.7. Using the "ddt" debugger to retrieve the symbolic and octal contents of the runable version (a.out) of the program in Fig. 4.5.

TABLE 4.3

"ddt" COMMAND HIGHLIGHTS

(1) To access "ddt":

% ddt [debugs a.out]

or

% ddt filename

(2) Command structure:

[⟨address expression⟩ [, repetition]]⟨command⟩[⟨display mode⟩]

Items enclosed in brackets are optional.

(3) Commands: (partial list: see DDT(I), the UNIX system manual)

/	open address and display contents
=	evaluate and display address expression
!	store the address into the last specified memory location
new line	displays next memory location
;r	prints registers
n;s	single step; n = # of steps
;g	go at location 0
n;b	set breakpoint at n
x;c	clear breakpoint at x
;f	exit (control d exits also)
;d	display breakponts
;e	display floating registers

(4) Display commands:

i	display instructions
o	display octal values

For example, Print lines 0–5 in symbolic code

0,5/i

Print lines 0–5 in octal

0,5/o

```
% nm test
000110t buffer
000134t count
000032t get
000076t k
000010t loop
000000t start
```

Fig. 4.8. Symbol table generated by using the "nm" command on the a.out file generated from the code in Fig. 4.5.

```
% as  test.as
% mv  a.out test
% test
00017%
```

Fig. 4.9. The standard sequence used for creating a running program from 'as' code.

```
% ddt test

016;b
;g
016     bpl     010     ;r
r0                      01          01
r1                      0100        0100
r2                      0           0
r3                      0           0
r4                      0           0
r5                      0           0
sp                      0177764     0177764
pc                      016         016
;p
016     bpl     010     ;r
r0                      03          03
r1                      0102        k+04
r2                      0           0
r3                      0           0
r4                      0           0
r5                      0           0
sp                      0177764     0177764
pc                      016         016
;p
016     bpl     010     ;r
r0                      06          06
r1                      0104        k+06
r2                      0           0
r3                      0           0
r4                      0           0
r5                      0           0
sp                      0177764     0177764
pc                      016         016
;p
016     bpl     010     ;r
r0                      012         012
r1                      0106        k+010
r2                      0           0
r3                      0           0
r4                      0           0
r5                      0           0
sp                      0177764     0177764
pc                      016         016
;p
016     bpl     010     ;r
r0                      017         017
r1                      0110        buffer
r2                      0           0
r3                      0           0
r4                      0           0
r5                      0           0
sp                      0177764     0177764
pc                      016         016
;p
00017Process terminated.
```

Fig. 4.10. A "ddt" run on the program "test," generated from the code in Fig. 4.5.

to run the program from 0 to location 16 and pause. "016" is in the loop, so pausing (or "breaking") at 16 will allow us to see the addition occur. Next, type ;r to print the registers.

Figure 4.10 shows the results of issuing the above commands. As expected, r0 = 1 at the first break. Now by typing ';p' to proceed, the program will loop back and break once more at 16 with r0 = 3. Continuing this procedure, r0 will contain 017 (the answer) after five iterations. By issuing a ';c' breakpoints may be cleared. A ';s' enables single stepping (executing one instruction at a time).

4.6 'as' PROCEDURE SUMMARY

Several common procedures for using 'as' are summarized as follows:

% make prog.s $ex$$	[create ASCII file named prog.s]
% as prog.s – la	[assembles file and searches /lib/liba.a for library routines, e.g., sine, etc.]
% a.out	[runs program]
% mv a.out prog	[moves a.out to permanent filename]
% nm prog	[tells where symbolic names are located] [will not work if previously stripped]
% strip prog	[removes symbol table to get smaller file]
% ar t /lib/liba.a	[will print what is in library]

The reader is encouraged to read appropriate sections of the UNIX system manual to understand each command listed above.

A variety of debuggers other than "ddt" are available. There is "cdb" (the 'c' debugger) and "db," for example. "cdb" and "db" are part of the original UNIX system version 6 distribution. "ddt" was included in this text as an example because it was found, after testing various other debuggers, to have many useful features. "ddt" was obtained from the UNIX user's group distribution at the Rockefeller University.

4.7 SYSTEM CALLS

Section II of the UNIX Programmer's Manual contains the system calls that are available to the 'C' and 'as' programmer. Table 5.4 in Chapter 5 shows some commonly used calls available in version 6 of the UNIX system for both 'as' and 'C'.

System calls are accessed by including the call name directly in the code. For example,

```
mov $1, r0
    .
    .
    .
sys exit
```

causes an exit to the UNIX system. In 'as' the mnemonic 'sys' signals the assembler to machine code a trap instruction (0104400) or'ed with the code for the system calls. Each call has a specific code number. For example, exit has the code 1. Thus, sys exit will actually become 104401 or TRAP 1. The software trap provides a way to use operating system facilities. The value in the last 6 bits of the trap instruction can be decoded by the operating system to determine which call action should be taken. Note that the Macro call in Fig. 4.2, listed as TRAP 1, provides the same exiting ability as sys exit in 'as'.

In cases in which variables are to be supplied to a system call, variables follow the call. For example, to write to a file three words are used:

```
            .
            .
            .
sys write
buffer address in memory
number of bytes to write
            .
            .
            .
```

A specific example of the use of this convention was shown in Fig. 4.5, where the instructions

```
        mov $1, r0
        sys write
        buffer
        5
          .
          .
          .
buffer: .even
        .=.+20.
```

```
start:  mov $0,r0           /the comments in fig4.5 are applicable
                            /to this program with the addition
        mov $k,r1           /of the comments shown below
loop:   add (r1)+,r0
        cmp $k+10,r1
        bpl loop
        mov $5,count
        mov $buffer+5,r3
get:    mov r0,r4
        bic $177770,r4
        add $60,r4
        movb r4,-(r3)
        asr r0;asr r0;asr r0;
        dec count
        bne get             /when a file is created,
        sys creat           /the file descriptor is returned in r0
        junk                /pointer to name of the file in ascii
        0666                /mode permissions - see chmod
        sys write           /the file descriptor created in
                            /'sys creat' is used
        buffer
        5
        sys close           /close the open file
        sys exit
k:      1
        2
        3
        4
        5
buffer:
        .even
        .=.+512.


count:  0
junk:   <result>            /the file name is result.
                            /The <...> enclose an ascii string.
```

Fig. 4.11. A modified version of Fig. 4.5. The ability to write the result to a file is included.

were used to write out values contained after the label "buffer." "r0" contains the file identifier. The value "1" put in r0 indicates writing on the standard output. "0" is the standard input, and "2" is an error output. Zero and one refer to your terminal. Files, such as those on the disk can be assigned numbers by a "sys open" call, which returns the file descriptor in r0.

The sequence of calls to read or write to and from files include the following calls:

creat	create a file
open	open an existing file for reading or writing
read	read a file
write	write on a file
close	close the file

Before using these calls, the user should study the information in Section II of the UNIX Programmer's Manual.

Figure 4.11 shows a modification of the program in Fig. 4.5 to allow writing on a file using some of the above system calls. First, a file named "result" is created in mode 666 (i.e., read, write on file is allowed. See UNIX system manuals and Chapter 5). Then, the 5-byte result is written and the file closed. Using read is similar to write. Note the use of <result> to specify the file name at label junk:. The file "result" may be examined using "od." For example, " % od result," will print the contents of the file.

In 'C', system calls are accessed by function calls. For example,

```
creat(. . .args. . .);
read (. . .args. . .);
write(. . .args. . .);
open(. . .args. . .);
close(. . .args. . .);
```

The use of calls in 'C' will be discussed in the next chapter.

EXERCISES

1. Write a program to print the character string

"This is a test"

(a) Write the program in 'as'

(1) Storing the characters in an array that you set up;
(2) Storing the characters using the 'as' facility to convert characters between angle brackets into a string of ASCII values.

(b) Write the program in 'Macro'

(1) using the Macro facility;
(2) not using the Macro facility.

After the string, be sure to insert a new line. For this problem, use all the UNIX system I/O facilities. Chapter 6 will address problems concerned with conducting these same operations on a standalone 11/03 without the UNIX system calls.

2. Write a program for adding numbers that you type in, e.g.,

```
% add
5
5
10
```

   (a) Read the ASCII numbers in using the "read" system call.
   (b) Convert to integer.
   (c) Add.
   (d) Write out using the write system call.
   (e) Consider how you would conduct the above operations in either decimal or octal. Explain.
   (f) How would you construct a program that would:

      (1) Accept positive and negative numbers?
      (2) Any length number?
      (3) Recognize standard syntax, i.e., 1+2−1+4−3= . . . ?

3. Write a program with a loop that iterates many times, e.g.,

```
       mov #2000, r3
loop:  tst -(r3)
       bne loop
       .
       .
       .
```

   Compute the execution time of this loop:

   (a) By hand, using timing information for the instructions.
   (b) As a test program using

```
% time test
```

   in the UNIX system.
   (c) Use "ddt" to run this program. If you allow "ddt" to iterate for 2000 times using the breakpoint facility, what happens? Why?

4. Show that "TRAP 1" in "Macro" and sys exit in 'as' are the same.
5. In Fig. 4.7, in location 070 there is a statement "jmp *r0." Explain this statement.
6. After assembling a test program, use "strip" to strip off the symbol table. How much space do you save?

5

Software Concepts—An Introductory User's Guide for the UNIX* Operating System and 'C' Programming Language

5.1 INTRODUCTION

This chapter presents the fundamental methods for operation of minicomputer systems that employ the UNIX operating system and the 'C' programming language. Four articles provided with the UNIX system documentation should be read first by the novice UNIX system user.

1. "UNIX for Beginners" by B. W. Kernighan
2. "Programming in C—A Tutorial" by B. W. Kernighan
3. "C Reference Manual" by D. M. Ritchie
4. "The UNIX Time-Sharing System" by D. M. Ritchie and K. Thompson

These four articles are provided with the sixth version of the UNIX system. A seventh version of the UNIX system is now available from Bell Laboratories with new features primarily advantageous for larger PDP-11s (e.g., 11/70s). As described in the fourth article listed above, the UNIX system is a time-sharing system developed at Bell Laboratories and currently in heavy use throughout the Bell System and in many universities in the United States as well as in private industry. This operating

* UNIX is a trademark of Bell Laboratories.

system is made available to universities for only a small handling charge. It has become extremely attractive to universities, not only because it is inexpensive but also because it is a superb, easy-to-use, and almost crash-free system. Further, the system, written in programming language 'C', is provided to the user in source form. Thus, the system may be readily modified. Many university systems are modified from the original. Some simple modifications can be made to enable systems to have attributes somewhat more like commercial systems supplied by Digital Equipment Corporation (RT-11, RSTS, RSX). For example, the rubout command is often modified to be a "rubout" character rather than the # sign provided in original the UNIX system. Likewise, it is desirable to allow ^C (control c: press control and c simultaneously) to cause exiting from a program.

This chapter will discuss (1) some of the most important ideas described in the "UNIX for Beginners" document, and (2) several items in the 'C' tutorial and reference manual. Document #4 in the above reading list is included to provide the reader with a more technical overview of the system. Readers unfamiliar with the UNIX system should scan this paper but not initially try to understand it in depth. Also, the 'C' reference manual may not be needed since the same information is contained in the 'C' tutorial. More detailed information and extensive examples can be found in a book entitled "The C Programming Language" by Kernighan and Ritchie, Prentice-Hall, 1978.

5.2 UNIX AND 'C' DOCUMENTATION

There are only a small number of documents available to the UNIX and 'C' systems user. All are useful and will help both novice and experienced users deal with the problems that arise. These documents are

1. The UNIX Programmer's Manual, K. Thompson and D. M. Ritchie, 1975.
2. A collection of UNIX documents supplied with the UNIX system.

   (a) Setting up UNIX—Sixth Edition
   (b) The UNIX Time-Sharing System
   (c) C Reference Manual
   (d) Programming in C—A Tutorial
   (e) UNIX Assembler Reference Manual
   (f) A Tutorial Introduction to the UNIX Text Editor
   (g) UNIX for Beginners

(h) RATFOR—Rational Fortran
(i) YACC—Yet Another Compiler Compiler
(j) NROFF User's Manual
(k) The UNIX I/O System
(l) A Manual for the Tmg Compiler Writing Language
(m) On the Security of the UNIX System
(n) The M6 Macro Processor
(o) A System for Typesetting Math
(p) DC—Desk Calculator
(q) BC—Desk Calculator
(r) The Portable C Library
(s) UNIX Summary

3. Bell System Technical Journal, "UNIX Time-Sharing System," Vol. 57, July–August, 1978.
4. "The C Programming Language" by Kernighan and Ritchie, Prentice-Hall, 1978.
5. "UNIX Operating System Source Code Level 6," J. Lions, University of New South Wales, 1977
6. "A Commentary on the UNIX Operating System," J. Lions, University of New South Wales, 1977.

The documents listed above should be part of every UNIX system user's library. The two main manuals, (1) and (2), are supplied with the UNIX Operating System obtained from Bell Laboratories. Document (3), also available from Bell Laboratories, is a useful commentary on the UNIX system. "The C Programming Language" is an excellent inexpensive paperback text that describes the C programming language in detail. The companion volumes (5) and (6), by Dr. J. Lions in the Department of Computer Science at the University of New South Wales, Australia, are available from Bell Laboratories to holders of a UNIX system license.

5.3 OBTAINING A UNIX SYSTEM LICENSE

Universities can obtain a license for the UNIX system from

Computing Information Service
Bell Laboratories
600 Mountain Ave.
Murray Hill, N.J. 07974
(201) 582-3000

for a nominal charge. A license is for a single CPU. Additional use of UNIX system on other CPUs must be approved by Bell. Similarly, the

MINIUNIX* system can be obtained from the Bell System. There is a UNIX system user's group that holds semiannual meetings. A UNIX tape distribution exists, which contains many useful programs. This distribution was originally provided to users on magnetic tape by The Rockefeller University.

5.4 PROGRAMMER'S MANUAL

The UNIX operating system supplied to universities comes with a programmer's manual that explains the commands available to the user and other system-related commands. To learn effectively how to use the UNIX system, the student should first read the manual and, second, refer to it frequently while using the system. There are eight parts to the manual:

1. Commands—this is the most-used section
2. System calls
3. Subroutines
4. Special files
5. File formats and conventions
6. User-maintained programs
7. User-maintained subroutines
8. System maintenance

The first three parts are normally sufficient for most users. Part 1 summarizes all commands and part 2 describes system calls (i.e., ways to use the operating system facilities directly in programs). Part 3 contains subroutines (e.g., sine, cosine, etc.) that are often used. Part 4 on special files describes the files for specific types of hardware, while part 5 discusses how files are formatted and other conventions. The programs described in part 6 are often deleted from many systems, since the predominant program types in this section are game programs. The remaining sections contain other useful but not essential information for the beginner.

The novice UNIX system user is urged to first read the introductory material in the programmer's manual followed by a careful reading of the available commands.

5.5 USE OF UNIX SYSTEM IN THE LABORATORY

The most popular use of the UNIX system is on relatively large PDP-11 machines (e.g., 11/34, 11/70 with memory sizes of 64–128K words of

* MINIUNIX is a trademark of Bell Laboratories.

memory), where up to about 10–20 users can efficiently share the resources of the machine. On machines without memory management (limited to 32K words or less), an adaptation known as the "MINIUNIX" system is provided by Bell Laboratories that will run on PDP-11/04s up to PDP-11/40s. The MINIUNIX system is easily modified to run on LSI-11 microcomputer-based systems. One adaptation is known as LSX (BSTJ, July 1978) or MicroUNIX. The MINIUNIX system was initially configured to run with a PDP-11/40. The modifications necessary to run the MINIUNIX system on the 11/03 are (1) changing instructions that access the switch register on the 11/40 (there is no switch register on the 11/03), and (2) making use of the MTPS and MFPS (move to/from processor status) instructions on the 03 rather than the direct dumping of values for the PS into the 11/40 highmemory locations that are allocated to the PS. Appendix B gives a brief description of the changes necessary to convert the MINIUNIX system to the MicroUNIX system.

For use in a laboratory environment, we have employed UNIX system on a central system connected to remote LSI-11-based systems that either run (1) the MicroUNIX system, (2) standalone, or (3) RT-11 (DEC). Figure 5.1 shows the configuration of the general laboratory system used for generating examples in this text. Our experience has been that such a system is generally more useful than a MicroUNIX system. MINI- or MicroUNIX systems running on floppy disks tends to be very slow and not entirely bug-free. On the other hand, the UNIX system using memory management is essentially bug-free and crash-proof. Combining a raw machine (11/03) with a versatile operating system (the UNIX system on host computer) for producing the code on the small machine results in the happy circumstance of efficient program operation and a low-cost small system for dedicated laboratory use. Similar systems or systems with similar principles can be easily generated in any laboratory with PDP-11s.

A fundamental concept for laboratory computing is that real-time responsiveness is required. This requirement necessitates use of a small computer at the experimental site. However, most small facilities provide limited disk space, compile programs slowly, and generally do not perform as well as larger systems. Therefore, the concept of the distributed laboratory central system with one larger system for serving several smaller machines is viable. Consider how a user with a minicomputer might use a system. A remote user with a small computer could be physically separated from the host and communicate via telephone lines or be directly linked with a serial or parallel line. The scenario proposed is to use a relatively large minicomputer (e.g., an 11/34 or 11/70) for compilation and file storage for a number of PDP-11/03s with no mass storage.

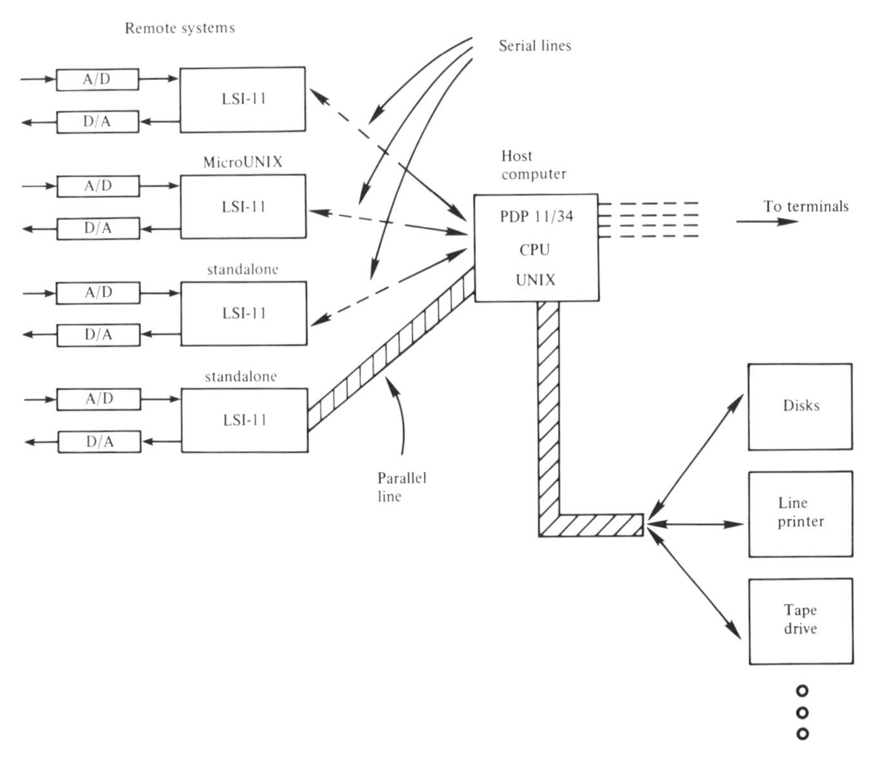

Fig. 5.1. A laboratory system using the UNIX operating system on a PDP-11/34 and RT-11, MINIUNIX, or no-operating system are used on the remote LSI-11s.

Each 11/03 might contain only a CPU, memory, analog-to-digital, and digital-to-analog conversion and clock (timing) facilities. Users would store and compile programs on the 11/34 and transfer the programs to the 11/03 for running. This facility has been implemented at Vanderbilt University and works very well. Two configurations are available: (1) one system uses a read-only memory for programs that communicate with the 11/34, and (2) the second system uses an RT-11-based system for storage of programs on a floppy disk. The operation of the current system will serve to illustrate the concept:

On the 11/03:

```
.R VTTY          [Run serial communication program on RT-11]
                 [VT – means virtual terminal]
login: john      [We are now in the UNIX system]
%                [(compile program here)]
```

```
% out-sav a.out      [produces a format suitable for loading]
                     [a file named a.sav with RT-11]
% to03vt a.sav       [moves file to 11/03]
.R 03.DAT            [Run program on 11/03]
```

Clearly, any type of system may be used for transferring data between computers. Both parallel and serial methods will work; the convention described above is simply one implementation. Another that has been used is to place the transfer program in ROM. In this case odt microcode on the 11/03 is used to start the transfer program (e.g., 140000G). The notion of converting headers (i.e., changing a.out to a.sav to run on 11/03) is useful only when different loading facilities are to be used. Perhaps the most straightforward implementation is to read each byte from the serial or parallel line and place bytes directly into memory.

Appendix C contains a list of sample programs that can be used to implement the transfer programs referenced above. A program written in 'C' for parallel transfer between an 11/03 and 11/34 using DRV-11 and DR-11C parallel cards is given. A similar implementation for serial transfer is also listed using RT-11 as a host-operating system on the 11/03. Individuals interested in bringing up a system for their own use can study this example and adapt it to their own needs. Macro was used for assembly of vtty.mac to produce code running on an 11/03 with floppy disks. A floppy-disk-based 11/03 system is probably the most common configuration found in laboratories today. 'as' or 'C' could be used, as well, to produce code for the transfer program. The author recommends producing the code in C and storing in ROM for use on the 03.

The majority of the work discussed in this chapter and the next can be performed on a UNIX time-sharing system without an 11/03. The multi-CPU systems described above will not be used until Chapters 6–9. Moreover, the information presented on laboratory usage of minicomputers can be fruitfully studied even without the system described. For example, a small disk-based 11/03 system can run the programs described using an operating system such as RT-11 or the MINIUNIX system. Further, 'C' can be used with DEC systems, such as RT-11 or RSX, by using compilers provided by Whitesmiths Ltd. in New York City. For persons not having access to a UNIX system but still wanting to use 'C' for laboratory minicomputing, employing a Whitesmiths' compiler running on another operating system will be a quite efficient way to proceed. A standalone system that would work quite well could be an 11/03 (or 11/23) running RT-11 with 2-RL01 disk units. The compiler for RT-11 would allow programming in the 'C' language with all its attendant advantages and allow simple programming of peripherals.

5.6 USING THE UNIX SYSTEM

First, read "UNIX for Beginners." One common modification is that the sharp-character "#" is frequently changed to a "rubout" character and the "@" character is changed to a control U(^U) (erase entire line). These characters can be selected by the user with an 'stty' command, as well. This article discusses the UNIX editor 'ed.' "teco" is recommended for general use rather than 'ed'. See Appendix C for information on obtaining UNIX-based teco. Teco is a character-oriented editor and is in widespread use on many different DEC computers and DEC-operating systems. There are, however, reasons for using 'ed' occasionally, so it may be useful to scan this section of the paper. The remainder of the document is clear. You may not want to read Section III on document preparation, unless you wish to prepare documents.

Example. Creating a file:

Follow the steps below to create a file with teco. If you are using another editor, create the same files with that editor.

% make junk (underline denotes characters typed by the UNIX operating system)
* i This is a test filex CR (carriage return)
$ex$$ ($= escape character, ex= exit)
%

"make" is a command on some systems that calls teco and allows you to insert characters. "i" is the command for insertion, i.e.,

i string$$

causes the string (however long it is) to be put in the text buffer. If now you wish to edit the file,

% teco junk
*^^'ev$$ (lets a pointer "'" to characters be displayed; type control^, the desired pointer character, ev and two escapes)
* fsfilex$file$$

will change "filex" to "file."

*ex$$
%

See Table 5.1 for a summary set of teco commands. It is recommended that all newcomers to teco stop here and practice with teco, trying the examples in the Teco User's Manual.

TABLE 5.1

SUMMARY OF FREQUENTLY USED teco COMMANDS

teco ⟨file name⟩	To edit preexisting file
make ⟨file name⟩	To create new file under teco
l	Move pointer to start of following line
v 'verify'	Prints out line
t 'type'	Prints line starting at pointer to end of line
0l	Move pointer to beginning of line
0t	Type from beginning of line to pointer
j	Move pointer to start of buffer
zj	Move pointer to end of buffer
i ⟨text⟩ $	Insert text at pointer location
d	Delete character following pointer
c	Move pointer over one character to the right
p 'page'	Take buffer and write it out to output file and get next page from input file
s ⟨str⟩ $	Search string. When located, pointer is after last character in string.
fs[old string] $[replacement string] $$	Finds the character string in the text and replaces it with a new string

Next, let us try a few UNIX commands. "junk" is the file you created. If you type

% cat junk

'This is a test' will appear on your terminal. If you type

% number junk

the same phrase will be printed but there will be a sequential number for each line in the leftmost column. This is a handy command since program diagnostics refer to line numbers. If 'number' is not on your system, it can be easily added.

Further, consider the redirection facilities of the UNIX system. You can, for example,

% cat junk >/dev/la1

or

% cat junk >/dev/tty2

In the first case, your file will be printed on the line printer and in the second case on the terminal (#2 in this case). The '>' is a redirection symbol; it feeds the standard output (your tty) somewhere else. A file in directory /dev (stands for device) has a name la1 or tty2 (see next section for file structure definitions), which permits output to be directed to these devices. Devices, e.g., terminals, are accessed through special files, each having a unique name (see Section IV of your programmer's manual). If you redirect your output to that name, it will appear on that device.

5.7 FILE ORGANIZATION

The UNIX file system is organized like an upside down tree. Figure 5.2 displays the structure of a typical file system. The root (top or beginning) of the system can be accessed by use of the / symbol. 'cd' is the same as "chdir," which means change directory. It is interesting to explore a bit. Type "ls −1" and see what appears. Figure 5.3 shows an example. Try also ls −1 /usr to see who is a user on the system. Read page 6 of the "UNIX for Beginners" paper if you are confused.

Several important directories are children of the root. 'bin' contains a set of run files that are scanned whenever a user issues a command (for example, "ls"). The program for doing an "ls' is found in bin. If any program is run frequently by most users, the run file can be put in the bin directory for easy access. If too many people do this, however, bin tends to get rather cluttered. It is better to create another directory just for user commands, e.g., "ubin." "UNIX" contains UNIX files, often different

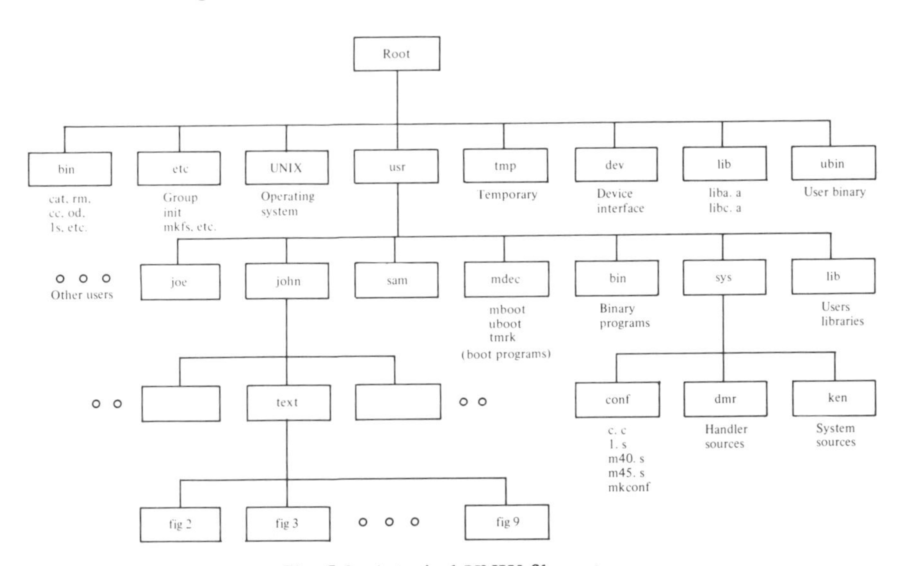

Fig. 5.2. A typical UNIX file system.

After logging on, typing "ls -l" will, for example, produce:

```
total 10
drwxrwxrwx  2 john       64 Feb 25 14:41 bin
drwxrwxrwx  2 john      160 Feb 25 14:40 libe
drwxrwxrwx  2 john      288 Feb 25 14:05 misc
drwxrwxrwx  2 john      128 Sep  9 12:25 sampling
drwxrwxrwx  2 john      176 Feb 20 16:46 smarti
drwxrwxrwx  2 john      336 Feb 25 14:06 stat
drwxrwxrwx  2 john       48 Sep  9 12:25 subop
drwxrwxrwx 10 john      752 Feb 26 11:01 text
drwxrwxrwx  2 john      304 Jan 15 09:42 ttytest
```

Then changing directories: i.e. % cd text and % ls -l produces:

```
total 8
drwxrwxrwx  2 john       96 Feb 21 14:21 appendix
drwxrwxrwx  2 john       80 Feb 20 10:29 fig2
drwxrwxrwx  2 john      256 Feb 25 07:54 fig4
drwxrwxrwx  2 john      160 Feb 20 10:47 fig5
drwxrwxrwx  2 john      224 Feb 21 11:19 fig6
drwxrwxrwx  2 john      192 Feb 25 14:02 fig7
drwxrwxrwx  2 john      384 Feb 20 10:38 fig8
drwxrwxrwx  2 john      400 Feb 26 10:48 fig9
```

Then, at the lowest level: cd fig8 and ls -l:

```
total 43
-rw-r--r--  1 john     2436 Feb 20 10:30 callC.as
-r--r--r--  1 john      621 Feb 20 10:38 clkintr.c
-r--r--r--  1 john     1556 Feb 20 10:37 echo_int.c
-rw-r--r--  1 john      113 Feb 20 10:36 edjob
-r--r--r--  1 john      975 Feb 18 09:20 fig8.11
-rw-rw-r--  1 john     2093 Feb 18 09:20 fig8.12
-r--r--r--  1 john      621 Feb 18 09:20 fig8.13
-rw-rw-r--  1 john     1317 Feb 18 09:20 fig8.15
-rw-rw-rw-  1 john     1116 Feb 18 09:20 fig8.16
-rw-rw-rw-  1 john      349 Feb 18 09:20 fig8.17
-r--r--r--  1 john      895 Feb 18 09:20 fig8.19
-rw-rw-r--  1 john      514 Feb 18 09:20 fig8.20
-rw-rw-r--  1 john      808 Feb 18 09:20 fig8.23
-rw-r--r--  1 john      521 Feb 18 09:20 fig8.25
-r--r--r--  1 john     2446 Feb 18 09:20 fig8.8
-r--r--r--  1 john      929 Feb 18 09:20 fig8.9
```

Fig. 5.3. An example directory listing.

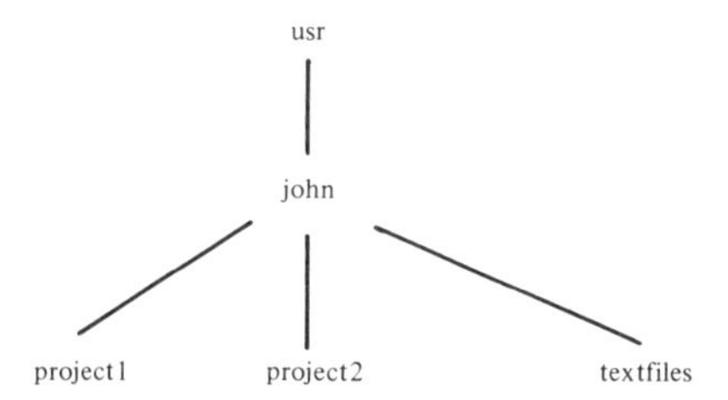

Fig. 5.4. An account tree of directories.

versions of the UNIX system. "tmp" is for temporary storage. 'dev' contains interface code to all the device handlers on the system. 'lib' is for libraries and 'etc' is a catch-all directory. 'usr' contains most of the other directories and programs in the file system.

The use of a tree structure for files allows the user to organize his own account. For example, a common strategy is to create in one's own account a set of directories: one for each set of items one is working on. When one logs onto the system, one is at the portion of the tree /usr/yourname (e.g., /usr/john; see Fig. 5.2). You can now create other directories. Directories are simply lists of files, other directories, or a mixture of both. The 'ls − l' command will tell you whether a named file is a directory or not [d (directory) or __ (not a directory) in the first column].

Let us make some directories:

```
% mkdir project1
% mkdir project2
% mkdir textfiles
```

The account tree now appears as shown in Fig. 5.4. Suppose we make a file in project2. First we change directories by

```
% cd project2
% make junk
*  Ithis is a test CR
   $ex$$
%
```

The tree now appears as shown in Fig. 5.5.

You can move from one directory to another by several methods: For example,

1. You can type

```
% cd /usr/john/project2
```

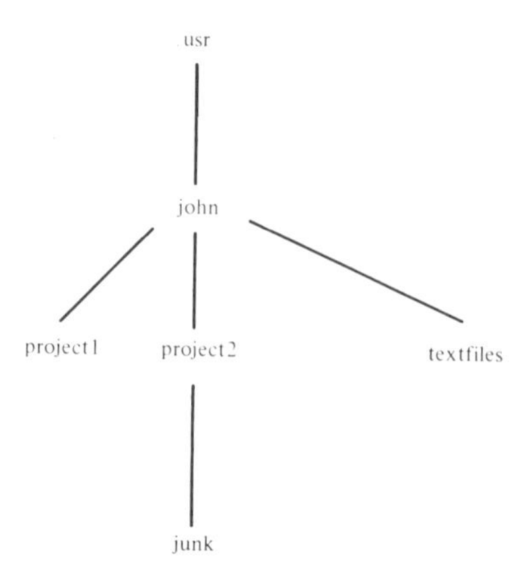

Fig. 5.5. Addition of a text file to "project 2" directory.

2. If you are in project2 you can type % cd . ., where ". ." refers to the most immediate ancestor (parent) of your current directory (jargon; just means you move up one node on the tree). Then type % cd project1. When you issue "cd . ." you moved to /usr/john, then by % cd project1 you moved down the tree again. Thus you must always specify the complete "path" (how you get there) name or move from point to point on the tree.

Listings. To see what is in your directory (or any other as well), type

% ls −1

The −1 is a flag that means long format. Figure 5.3 shows an example of the result of using this option with ls. Section I of the programmers manual [ls(I)] gives other alternatives and explains the information that is printed in Fig. 5.3.

5.8 PROGRAMMING IN 'C'

'C' is a structured programming language that allows users to write programs simply and quickly at a high level while retaining many of the attributes that make assembly code attractive. The reprint "Programming in C—A Tutorial" by Brian W. Kernighan is available in the UNIX system documentation. The novice 'C' user should read this article and try exercises that are suggested. It may be worthwhile to examine the "C Reference Manual," as well. However, most users will find most information they need in the tutorial. The text entitled "The C Programming Language" by Kernighan and Richie (1978), contains many useful examples and tutorial information.

5.8.1 Structured Programming

A primary virture of the use of 'C' is that it is a structured language; i.e., it provides a logical flow of coding that is readily and modularly related to the needs of the programmer. A simple example for sampling the A/D is

```
main ( ) {
          setup( );
          clock( );
          display( );
        }
```

Each function is called sequentially and each in turn can be composed of sets of functions. For example,

```
setup ( ) {
           definedata( );
           zerobuf( );
           setparams( );
                  .
                  .
                  .
         }
```

Thus each function can be made up of other functions that are defined by name. This type of arrangement makes it very easy to read and understand programs.

This next section will build on concepts described in the tutorial and step through a complete example giving attention to interaction with the UNIX operating system.

5.8.2 Creating and Running Programs

Perhaps the most straightforward program to write is a program to printout a string, e.g., "This is a test."

First, create the file

```
% make test.c
Imain( ){
                    printf("This is a test\n");
            }
$ex$$
%
```

Teco is first entered using "make" and the code shown inserted. "main( )" is the entry point of the program and printf is a standard system utility function that prints the string in quotes. "\n" means new line and return. Note that the line is followed by a ;. Each line must be terminated by a ';'. Statements may be broken into several lines with the last one terminated with a semicolon.

"cc" is the command to compile. A variety of options are available at compile time. Options are included by typing a character after a − sign, e.g., cc −c test.c.

If you type "cc test.c" no options are involved and the program is compiled and a file named "a.out" is produced. To run the program, one simply types a.out. A common procedure is to rename the file with some other name in order to save it, since on each compile a file named a.out will be produced, writing over any previous a.out file. For example,

```
% cc test.c
% a.out
  This is a test
% mv a.out test
% test
  This is a test
%
```

Consider several popular options:

1. cc −c test.c will produce a file named test.o. This is an object file that can be subsequently linked with other files. The concept is that if there are a number of files to be combined ("linked") into one program, it is unnecessary to compile them each time you wish to create a run file (i.e., an a.out file).

2. cc −S test.c will produce an assembly code version of the 'C' program named test.s. Note that the option is a capital S. The file produced will be test.s. Reading .s files is instructive since 'C' conventions not completely understood can be related to the machine code.

3. cc −O test.c will optimize the code for a combination of speed and size. It is interesting to examine the code produced in assembly code both

with and without the −O. To examine the code, the following steps would be required:

```
% cc −O −S test.c
% cat test.s >/dev/la1
% cc −S test.c
% cat test.s >/dev/la1
```

You now have two versions of the 'C'-generated PDP-11 code printed out. The differences between the optimized and nonoptimized versions will be examined later in this chapter.

You may combine options on a line:

```
% cc −c −O test.c
% cc −S −O test.c
```

if the options do not specify conflicting operations. Another important notion is that you may include object files (i.e., already compiled) directly in the cc command line as well as library searches. For example, cc −O test.c func.o sub.o −la. This command first compiles test.c and then loads it into memory followed by func.o, sub.o, and also searches the library liba.a. The name of the library is libx.a, where the command line option is −lx ('x' may be multicharacter).

5.9 THE LIBRARY FACILITY

The UNIX system provides the user the ability to maintain his own library of object files (.o). Public libraries are maintained in the /lib directory. A typical printout of library routines are shown in Fig. 5.6. This

```
-r--r--r--  1 danny      96 Sep  9 12:09 crt0.o
-r--r--r--  1 mini    14498 Sep  9 12:09 libA.a
-r--r--r--  1 mini    24928 Sep  9 12:09 libC.a
-r--r--r--  1 bin      9106 Nov 17 20:36 liba.a
-r--r--r--  1 bin     23238 Jan 17 00:14 libc.a
-rw-rw-rw-  1 eeg     98466 Jan 17 00:05 libe.a
-rw-rw-rw-  1 gary    35138 Jan 23 10:28 libg.a
-rw-rw-rw-  1 root    17950 Sep  9 12:09 libn.a
-r--r--r--  1 bin     54998 Nov 15 15:45 libp.a
-r--r--r--  1 ingres  21078 Sep  9 12:09 libq.a
-r--r--r--  1 bin      3530 Sep  9 12:09 liby.a
-r--r--r--  1 bin       436 Sep  9 12:09 mcrt0.o
```

Fig. 5.6. A "ls −1" listing of /lib in a UNIX system.

listing can be obtained by typing

% ls −1 /lib

Various libraries are maintained including functions for use with 'C' programs and routines for use with 'as'. Users may put commonly used routines in a library. Two libraries maintained for users are

liba.a
libc.a

where liba contains various subroutines that are listed in Section III of the Programmer's Manual. A list of some of these routines is given in Table 5.2. ''libc.a'' contains system calls for 'C' and 'as' programs. A more detailed discussion of these routines is given later in this chapter and are fully described in Section II of the Programmer's Manual.

If a user wishes to create his own library of routines in /lib, a set of object (.o) files would first be generated (using the −c flag in cc). Then, with ar(I) these .o files can be put into /lib. First, change directories to /lib. Decide to name your library libz.a (for example)

% ar r libz.a name . . .

You should create a library file in your directory and then move it to /lib.

TABLE 5.2

USEFUL SUBROUTINES AVAILABLE WITH VERSION 6 OF THE UNIX SYSTEM

abs	Absolute value
atan	Arctangent
atof	ASCII to floating conversion
atoi	ASCII to integer
crypt	Password encoding
exp	Exponential
floor, ceil	Largest, smallest integer
gamma	Log gamma
getc, getw	Get character, word
getchar	Read character
log	Natural log
printf	Formatted print
putc, putw	Put out char, word
putchar	Write character
rand	Random number generator
sin, cos	Sine, cosine
sqrt	Square root

Options available are

r replace or append named files
t print names for files in archive
x extract
u same as r except only files that are modified are replaced
d delete
v verbose, describes what is going on

After files are in an archive, the user can access all of them by, for example,

```
% cc test.c -lz
```

Next, a complete example will be given for a simple square root program that illustrates many of the features of 'C'.

5.10 TUTORIAL EXAMPLE

Common operations in 'C' programs include using functions, passing arguments, and interfacing to the shell command line. The following example was designed to explain the use of these common procedures.

The program "root.c", shown in Fig. 5.7, is a program for finding the integer square root of a number. The basic algorithm is simple. Successive odd numbers (1,3,5,. . .) are subtracted from the number you are finding the square root of (1 first, 3 next, etc.). The number of subtractions required such that the remainder is less than or equal to 0 is equal to the integer square root. For example, for the square root of 25

```
 25
 -1    subtraction 1
 --
 24
 -3    subtraction 2
 --
 21
 -5    subtraction 3
 --
 16
 -7    subtraction 4
 --
  9
 -9    subtraction 5 = integer square root
 --
  0
```

Now consider the program in detail.

```
 1
 2  main(argc, argv)
 3
 4          int argc;
 5          char *argv[];
 6  {
 7          int j, n,odd,count;
 8
 9          n=j = atoi(argv[1]);
10          count=0; odd = 1;
11
12          for(odd=1;( n= n-odd)>=0;odd = odd + 2){
13  /*      printf("n-odd= %d and odd= %d\n",n-odd,odd );*/
14          count++;
15          }
16          printf("the integer root of %d is %d \n",j,count)
17  }
18
19
20
21  atoi(s)
22          char s[];{
23          int i,n;
24          n=0;
25          for (i=0;s[i]>='0'&&s[i]<='9';i++)
26                  n = 10*n +s[i] - '0';
27                  return(n);
28  }
```

Fig. 5.7. An example program ("root.c") for illustrating concepts in 'C'. This program finds the integer square root.

Line 2: main(argc, argv). argv is an argument pointer used to point to values on the line that the user types in. argc is the number of arguments. For example,

% root 25

produces 5 on the screen. root is one argument and 25 is the other.

In order to determine the square root of 25, the program must be able to access the 25. The shell (user-to-operating system communications facility of the UNIX system) provides this ability as follows:

(a) argc is declared as an integer and passes the number of arguments to the program. For example, argc = 2 for root 25.

(b) argv is an array of pointers to pointers. The declaration of such an array is

char *argv [];

or

char **argv

which are the same. The usual way to read this nomenclature is to start at the right of the string and read left, i.e.,

char **argv means:

"argv points to pointers to characters" or diagrammatically:

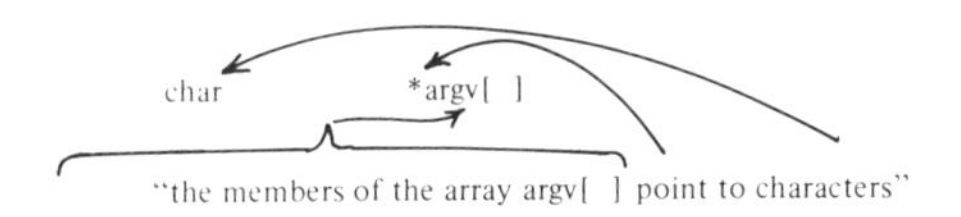

Consider the correspondence between argv[ ] and a command line (see Fig. 5.8). The actual values can be accessed via a double array, or the pointer to the value via a single array. That is, for "root 25": arv [1] is the pointer to 25, argv[0] is the pointer to the string "root" and argv[1][0] is the value 2, argv [1][1] is the value 5, and argv[1][2] = '\0'.

Lines 2–5: We declare the beginning of the program and the variable types necessary to pass arguments. Note that this declaration comes before entering the main braces "{" of the program found in line 6.

Line 7: Declare four variables as integers ($< 65{,}536 = 2^{16}$). Note that variables used in passing arguments are declared outside the braces (globally defined), while variables defined inside the braces are local to the routine they are in. These variables are called automatic and use stack locations for storage.

Line 9: The first problem encountered is that the value 25 you type in is in ASCII and should be converted to integer before arithmetic operations are performed on the number. Thus, argv[1][0] = 62 and argv [1][1] = 65, the ASCII equivalent of 2 and 5. By use of the function atoi we can

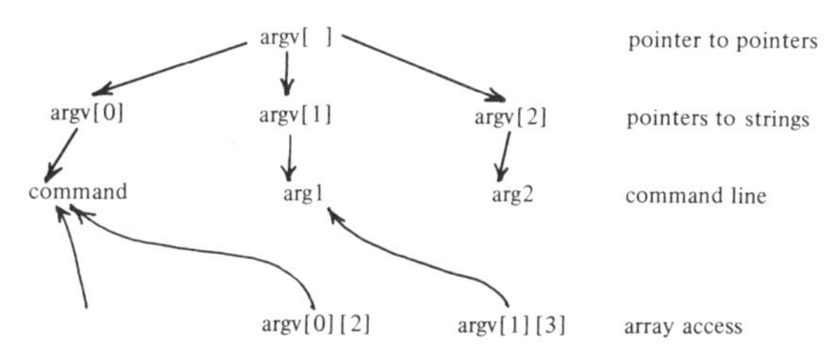

Fig. 5.8. Example of argv[ ] pointing to elements on a command line.

convert ASCII to integer. Thus, in line 9, we set n and j to the integer value for 25 (25 = 011001 base 2). Originally 25 is represented as

65	62

or, converting this octal byte representation to an octal word, we obtain

0 3 2 4 6 2

Thus, the ASCII word representation for 25 is 032462. "atoi" will convert 032462 to 000031_8, an integer representation.

Next, detour to lines 21–28 to see how atoi works.

Line 21: "s" is declared as an array of characters; therefore "s" points to the array of characters. When atoi(argv[1]) is called, argv[1] is a pointer to 62 (i.e., argv[1] is the value of the byte location that contains the value 62). Thus, s in the function atoi receives this address. When s is used in array form (s[1]), the actual values are retrieved. s[0] = 62, s[1] = 65. Note that *s would access the value 62 as well and ++*s would subsequently get 65.

Next examine the "for" loop in atoi. While the value passed is between ASCII 0 and 9, the statement

n = 10 * n + s[i] – '0'

is executed. For example, when s[0] = 62:

n = 10 * 0 + 62 – 60 = 2

In the next iteration, s[1] = 65 and n = 10 * 2 + 65 – 60 = 25 !! (the value).

Finally, the value n is returned: atoi in line 9 has the integer value 25 on return. Note that the n in main and in atoi are not connected in any way—they are 'automatic' variables, used only locally by pushing values on the stack.

One can also use the scanf function for accepting input rather than using the atoi function. For example,

```
scanf ("%d",&n);
j = n;
```

will return the same value as atoi. "scanf" is similar to printf except it is for reading rather than printing. Full documentation is found in "The C Programmer's Manual" and the "Portable C Library." The "Portable C Library" is usually contained in /lib/libp.a and thus scanned by the –lp flag in the command line.

Lines 12–15: The heart of the program. ''odd'' is the variable in which the values 1,3,5. . . are successively stored. ''odd'' is incremented by 2 on each pass. When the remainder n = n-odd is <= 0, the loop is exited. ''count'' contains the number of subtractions and is printed in line 16.

Line 13: A debugger line is commented out. A convenient way to debug programs is to put in printf statements everywhere. Here, when the /* . . . . */ are taken out, the values n-odd and odd are printed on each subtraction.

5.10.1 Running the Example Program

To create root.c you can type it in with teco or another editor:

```
% cc root.c
% mv a.out root
% root 25
5
% root 36
6
%
```

There are some interesting variations that illustrate redirection.

1. % make datafile
   i25
   $ex$$
   [the above creates a file containing 25]
2. next, create an empty file called
   ''resultfile.''
   [e.g., cat >resultfile, ^d]
3. % root <datafile >resultfile

will take data from datafile and place the result in resultfile.

5.11 ASSEMBLY LANGUAGE PRODUCED FROM 'C' CODE

Figure 5.9 displays the 'as' equivalent of root.c called root.s. One obtains root.s by

```
% cc -S root.c
```

There are many things to understand about the code generated by cc. First, there are data, text, and bss segments generated by the compiler.

```
.text
.globl   _main
_main:
~~main:
jsr      r5,csv
~j=177770
~n=177766
~odd=177764
~argc=4
~argv=6
~count=177762
sub      $10,sp
mov      6(r5),r0
mov      2(r0),(sp)
jsr      pc,*$_atoi
mov      r0,-10(r5)
mov      r0,-12(r5)
clr      -16(r5)
mov      $1,-14(r5)
mov      $1,-14(r5)
L2:mov   -12(r5),r0
sub      -14(r5),r0
mov      r0,-12(r5)
jlt      L3
inc      -16(r5)
L4:mov   -14(r5),r0
add      $2,r0
mov      r0,-14(r5)
jbr      L2
L3:mov   -16(r5),(sp)
mov      -10(r5),-(sp)
mov      $L5,-(sp)
jsr      pc,*$_printf
cmp      (sp)+,(sp)+
L1:jmp   cret
.text
.globl   _atoi
_atoi:
~~atoi:
jsr      r5,csv
~i=177770
~n=177766
~s=4
sub      $4,sp
clr      -12(r5)
clr      -10(r5)
L7:mov   4(r5),r0
add      -10(r5),r0
cmpb     $60,(r0)
jgt      L8
mov      4(r5),r0
add      -10(r5),r0
cmpb     $71,(r0)
jlt      L8
mov      -12(r5),r1
mul      $12,r1
```

```
mov      4(r5),r2
add      -10(r5),r2
movb     (r2),r2
add      r2,r1
add      $-60,r1
mov      r1,-12(r5)
L9:inc   -10(r5)
jbr      L7
L8:mov   -12(r5),r0
jbr      L6
L6:jmp   cret
.globl
.data
L5:.byte 164,150,145,40,
156,164,145,147,145,162,
162,157,157,164,40,157,
40,45,144,40,151,163,40,
45,144,40,12,0
```

Fig. 5.9. "root.s"—The 'as' equivalent of root.c.

"text" is the code, "data" is the initialized storage, and "bss" is uninitialized data storage. Many variables are "automatic"; that is, use the stack for temporary storage. For example, in root, j, n, odd and count are automatic. These variables appear defined at −10, −12, −14, and −16, respectively (i.e., 177770 = −10). These are location offsets from the stack pointer. At the very beginning, a "jsr r5,csv" is issued. 'csv' (c-Save) is a routine that pushes r2, r3, r4 onto the stack. When the cret routine is executed via a jmp cret (c return), these register values are restored. argc and argv are offset from the sp by 4 and 6, respectively, and r5 is filled with the address of the sp. Thus, in instruction number three after main,

```
mov 6(r5),r0
```

r5 + 6 refers to argv and moves the address of argv to r0.

```
mov 2(r0), (sp)    ; Stores argv[1] on the stack for the
                     following function call.
jsr pc, *$_atoi    ; Jump indirectly to the atoi function.
                     (note:_atoi in 'as' is the
                     equivalent of atoi( ) in 'C'.
mov r0, -10(r5)    ; Set j equal to the value
                     returned by atoi.
                     r5-10 refers to j.
mov r0, -12(r5)    ; Set n equal to the value
                     returned by atoi.
                     r5-12 refers to n.
clr -16(r5)        ; set count = 0.
mov $1, -14 (r5);  odd = 1.
```

The remaining statements may be similarly interpreted. Thus, code that does not work quite right for you in 'C' can be observed in an 'as' equivalent to see just what it is doing. Note that csv and cret are automatically included in programs. The two library options "−1c" and "−1" are also automatically added to the "cc" option list.

5.12 cret, csv, AND 'C' PROGRAM HEADERS

This section may be read now to aid in understanding what cret and csv functions do or it may be skipped and read in conjunction with Chapter 8. The code for cret and csv are not reproduced since the internal code of the UNIX system may not be disclosed to persons without a UNIX system license.

Basically, as noted earlier, csv saves registers and cret restores them after completing a function call. When a "jsr r5, csv" is issued, the contents of r5 are pushed onto the stack. Then, the value in r5 is stored; recall that r5 contains the address after the jsr (see Chapter 2). Next, the sp address is moved to r5 so that all automatic variables can be referenced throughout the remaining code relative to r5. r4, r3, and r2 are pushed onto the stack and finally a return is taken to the original calling routine. r4, r3, and r2 are stored at −2, −4, and −6 from the sp. Thus, the first automatic variable in the program will be stored at −10. In the root.c example, the first variable is j, because it was first listed in that "int j, ..." statement in root.c. The compiler assigns j the value −10; i.e.,

$$\tilde{}j = 177770$$

The other variables are then allocated space on the stack according to the order in which they were specified in the dimension statement. "cret" performs the reverse operation of csv, restoring the stack and returning.

The csv routine is part of the sequence of operations that are specified in a brief section of start-up code, generated for each 'C' program, that is contained in /lib/crt0.o. This code will vary in length depending on the needs of the user. For example, code to create an offset at the beginning of a program can be put in this file. An object file named /lib/crt20.o can be loaded using the −2 option in 'C'. Thus, one can retain one start-up file for use in some situations and another file for use in other situations. Basically, the start-up code is used to handle passing of arguments via the argc and argv variables in 'C'. When a program is entered, pointers to each string on the command line and the number of strings on the line are found on the stack. This small section of code determines the address of the arg list pointers and puts it on the stack too. Thus, the pointer to pointers of arguments is computed first in the start-up code. A jump to main occurs next. The important thing to note is that main usually starts at location 26. If you put a program in read-only memory (ROM), you can restart at

```
 1
 2  int       k[5] {
 3        1, 2, 3, 4, 5
 4  };
 5  main () {
 6        int      i,
 7                 sum;
 8        for (i = 0; i < 5; i++)
 9              sum = sum + k[i];
10        printf ("The sum is %d \n", sum);
11  }
```

Fig. 5.10. Program for adding five numbers.

```
 1
 2   int        k[] {
 3          1, 2, 3, 4, 5
 4   };
 5   main () {
 6          int        i,
 7                     sum;
 8          for (i = 0; i < 5; i++)
 9                sum = sum + k[i];
10          printf ("The sum is %d \n", sum);
11   }
```

Fig. 5.11. Replacing k[5] with k[ ] in Fig. 5.10.

program location +26 to access main directly. You may, however, need to set up the stack and argument passing facilities.

5.13 ADD EXAMPLE

This simple program, shown in Fig. 5.10, adds the numbers 1,2,3,4,5, and prints out the result 15. There are a number of possible modifications and experiments one can conduct with this program:

1. Line 2: int k[5] may be replaced with k[ ]. The compiler will fill in the number (see Fig. 5.11).
2. Line 9: Sum=+ k[i]; is shorthand for original statement:

sum = sum + k[i];

3. Pointers may be used. Add "int *p" to the variable list to define p as pointing to integers (see Fig. 5.12). The statement "sum=+ *p++"; may look unreadable, but is actually quite simple. If p is declared as a pointer,

```
 1
 2   int        k[] {
 3          1, 2, 3, 4, 5
 4   };
 5   main () {
 6          int        *p;
 7          int        i,
 8                     sum;
 9          p = k;
10          for (i = 0; i < 5; i++)
11                sum = sum + *p++;
12          printf ("The sum is%d\n", sum);
13   }
```

Fig. 5.12. Using a pointer to add five numbers.

setting p = k assigns p the address of the list k[ ]. "*p" means get the contents of the address in p. Thus, the first time we access p, the value 1 is returned. To get the other values in the list, we increment p on each pass through the loop, i.e., "*p++" where ++ after p indicates that the address should be incremented after it is used.

5.14 RUNNING THE ADD EXAMPLE PROGRAM WITH ddt

"ddt" is the dynamic debugging technique described in Chapter 4. There are a wide variety of operations supported by this program. The reader is advised to read the basic documentation for "ddt" before proceeding. "ddt" allows debugging of either 'C' or 'as' load modules (a.out form files). As an example of the use of 'ddt', let us examine the pointer program for adding five numbers (testpt.c). For convenience, name the load module testpt, i.e.,

```
% cc testpt.c   [produces a.out]
% mv a.out testpt
```

```
0
02
04
06
010         [start-up code from crt0.o]
014
020
022
024
026         jsr       r5,csv
032         sub       $6,sp
036         mov       $1326,p
044         clr       i
050         cmp       $5,i
056         ble       main+062
060         mov       *p,r0
064         add       sum,r0
070         mov       r0,sum
074         add       $2,p
main+054              inc        i
main+060              br         050
main+062              mov        sum,*sp
main+066              mov        $1340,-(sp)
main+072              jsr        pc,*$printf
main+076              tst        (sp)+
main+0100             jmp        cret
```

Fig. 5.13. Pointer version of add program in Section 5.13 obtained using ddt.

Figure 5.13 shows the results of using ddt to decode the file testpt. Locations 0–24 contain interface code to the operating system. This code is deleted from the figure since the code is part of the original UNIX system distribution. Interested readers with UNIX system can use ddt to examine the contents of these locations. Figure 5.14 shows the actual code (locations 26–76 in octal) generated and Fig. 5.15 shows the symbol table for the program (the latter is printed by "%nm testpt"). The symbol table may be removed from testpt by

% strip testpt

if you do not need it. Stripping the file will shorten it.

Examine Fig. 5.13 again. Main begins at location 26. As indicated in Section 5.12, the instructions before this location set up the argument pointers on the stack. This header information is obtained from the file /lib/crt0.o and is automatically added by cc. If another header is desired, it

```
0
02
04
06
010        [start-up code from crt0.o]
012
014
016
020
022
024
026        04567
030        01242
032        0162706
034        06
036        012765
040        01326
042        0177770
044        05065
046        0177766
050        022765
052        05
054        0177766
056        03414
060        017500
062        0177770
064        066500
066        0177764
070        010065
072        0177764
074        062765
076        02
```

Fig. 5.14. Octal version of Fig. 5.13.

```
001472D _fout
001326D _k
000026T _main
000132T _printf
001236T _putchar
000566t charac
001310T cret
000000a crt0.o
001274T csv
001274a csv.o
000326t decimal
001224a ffltpr.o
000754t float
001502b formp
001140t gnum
000644t nex
177766a i
000640t lnex
000654t loct
000446t long
000342t longorun
000156t loop
001506b ndfnd
001510b ndigit
001462d nulstr
000660t octal
177770a p
001476b pad_char
001512b pchar
001224T pfloat
001032t prbuf
000132a printf.o
001042t prstr
001224T pscien
001236a putchr.o
001020t remote
001504b rjust
001474B savr5
001002t scien
000000t start
000604t string
177764a sum
001356d swtab
000402t unsigned
001500b width
000026a x.o
000026t ~main
```

Fig. 5.15. Symbol table associated with pointer version of add program.

can be loaded using cc −2 . . . to load /lib/crt20.o. The user can modify this routine to add space, etc., as desired.

"p" receives the address 1326, which is the actual location of the list k relative to zero. "1326" is a considerable distance from 0, where the program started. The variables are stored after all functions are loaded; in this case printf consumes a large amount of memory past main+100. k[ ] is defined as an external variable list (not automatic, not on stack), and thus resides in memory in the data segment, after the text segment has been loaded. Note the difference where data storage is allocated: If the variable is defined outside main, storage is allocated permanently, but inside the function, storage is allocated on the stack. It is interesting to compare the listing in Fig. 5.13 with the .s versions in Figs. 5.26 and 5.27 in the problem set. These two renditions are essentially the same; it is just the way you read them.

The instructions that actually add the five numbers are

Location	*Instruction*
60	mov *p, r0
64	add sum, r0
70	mov r0, sum

```
% ddt testpt

 064;b
;g
064      add     sum,r0   ;r
r0                        01                   01
r1                        0                    0
r2                        0                    0
r3                        0                    0
r4                        0                    0
r5                        0177754              0177754
sp                        0177736              sum+02
pc                        064                  064
;p
064      add     sum,r0   ;r
r0                        02                   02
r1                        0                    0
r2                        0                    0
r3                        0                    0
r4                        0                    0
r5                        0177754              0177754
sp                        0177736              sum+02
pc                        064                  064
;p
064      add     sum,r0   ;r
r0                        03                   03
r1                        0                    0
r2                        0                    0
r3                        0                    0
r4                        0                    0
r5                        0177754              0177754
sp                        0177736              sum+02
pc                        064                  064
;p
064      add     sum,r0   ;r
r0                        04                   04
r1                        0                    0
r2                        0                    0
r3                        0                    0
r4                        0                    0
r5                        0177754              0177754
sp                        0177736              sum+02
pc                        064                  064
@sum=
06       ;p
064      add     sum,r0   ;p
The sum is15
Process terminated.
```

Fig. 5.16. Output obtained using ddt on add program.

Thus r0 gets the contents of p and sum is added to r0. Finally, the value is replaced in sum. By setting a breakpoint at location 64, we can run the program and pause (or break) at location 60 to see what has happened on each iteration of the add loop. Figure 5.16 displays the use of ddt in running this test.

```
064;b sets breakpoint
    ;g runs program starting at 0
    ;p proceeds
    ;r prints the value of the registers
    @sum prints what is in sum.
```

A program can be debugged using this technique, setting breakpoints inside loops and dynamically examining the contents of the registers at each breakpoint. In fact, this is a rather exhausting and tedious method for debugging 'C' programs. It is much easier to put in many printf statements to trace the flow of information in the program.

5.15 LONG INTEGERS

The use of the variable declaration

```
int i;
```

defines i to be an integer variable and permits the variable to assume only value from 0 to 65,535; that is, any value representable in a 16-bit word (2^{16} = 65,536). In some cases, it is desirable to use double-precision integers to form a 32-bit representation. If 32 bits are employed using the definition in 'C' of

```
long int i;
```

or equivalently

```
long i;
```

2^{32} = 4,284,967,296 values can be represented. In cases for which it is not desirable to use floating point (e.g., when speed is of importance) or for which floating-point hardware is not available, the use of "long" may be an appropriate strategy. Consider the example shown in Fig. 5.17 in which a variable "sum" is defined as long. "sum" is defined as "long" to avoid overflow in the summation. In Chapter 9 a specific example of this construction is given in a signal-averaging program.

```
 1
 2
 3  int i, x[500];
 4  long int sum;
 5  main()
 6  {
 7              sum = 0;
 8              for(i=0;i<500;i++)
 9                          sum =+ x[i];
10  }
11
12
13
14  .comm       _i,2
15  .comm       _x,1750
16  .comm       _sum,4
17  .text
18  .globl      _main
19  _main:
20  ~~main:
21  jsr         r5,csv
22  clr         _sum
23  clr         2+_sum
24  clr         _i
25  L2:cmp      $764,_i
26  jle         L3
27  mov         _i,r1
28  asl         r1
29  mov         _x(r1),r1
30  sxt         r0
31  add         r0,_sum
32  add         r1,2+_sum
33  adc         _sum
34  L4:inc      _i
35  jbr         L2
36  L3:L1:jmp           cret
37  .globl
38  .data
```

Fig. 5.17. 'as' code generated from 'c' code that uses a long integer.

Consider how the long integer is translated by the compiler into the .s code version. Four bytes are allocated for the double-precision sum:

comm _sum, 4

Additions to sum are carried out using the sign extend command "sxt" and the add with carry "adc" mnemonic. In line 29 the x[i] value is moved into r1. Then, r0 is set either to zero or −1, depending on whether the N bit was clear or set. Next, r0 and r1 are added to sum and sum+2, respectively. Finally, the carry bit is added to the most significant word in sum to complete the double-precision summation.

5.16 POINTERS AND STRUCTURES

Confusion often occurs among 'C' users who try to employ pointers with structures in their programs. Figure 5.18 illustrates some of the basic concepts for using pointers and structures and shows how the 'C' statements relate to machine code. The reader should read the section on use of structures in the 'C' tutorial or the text by Kernighan and Ritchie on 'C' before trying to understand the remainder of this section. Next, the details of this program will be explained line by line.

Lines 3–6: A structure is defined, which contains an integer x and a variable y that points to an integer. Name [2] declares that there are two sets of these variables with this structure and initializes p with the address of the structure (that is, p points to the structure). This is shown diagrammatically in Fig. 5.19.

The machine code (.s) equivalent file contains two lines, 22 and 23, that are equivalent to lines 3–6. ".comm _name, 10" allocates 10 bytes for storage of the variables. That is, x in name[0] uses 2 bytes, *y in name[0], 2 bytes, x in name[1], 2 bytes, etc., for total of $8_{10} = 10_8$ bytes. Two bytes are allocated following this block for p to store the address of the block.

Lines 10 and 29

```
p = name              (C)
mov $_name, _p        (as)
```

Setting p = name sets p to the address of name. Since name is an array, its address is taken when referred to this way. Note that p is correctly defined by using *p to indicate that p points to an integer. 'as' variables are the same as 'C' variables except they are preceeded by an _. When p is set equal to name (see Fig. 5.20), the address of name[0] is placed in location 10, since 2 bytes were allocated for *p after name[0] and name[1] (see line 23).

Lines 11 and 30

```
p -> x = 3;   (C)
mov $3, *_p   (as)
```

'p -> x' refers to x in the structure. Thus, x is set = 3, which is the same as moving 3 into x in 'as' (see Fig. 5.21).

```
 1
 2
 3  struct{
 4                  int x;
 5                  int *y;
 6          } name[2],*p;
 7
 8  main(){
 9
10          p = name;
11          p->x = 3;
12          *p->y = 4;
13          *p++->y = 5;
14          *(p++)->y = 6;
15          *(p++->y)++ = 7;
16  }
17
18
19  /The .s version of the above file is shown below
20
21
22  .comm   _name,10
23  .comm   _p,2
24  .text
25  .globl  _main
26  _main:
27  ~~main:
28  jsr     r5,csv
29  mov     $_name,_p
30  mov     $3,*_p
31  mov     _p,r0
32  mov     $4,*2(r0)
33  mov     _p,r0
34  mov     $5,*2(r0)
35  add     $4,_p
36  mov     _p,r0
37  mov     $6,*2(r0)
38  add     $4,_p
39  mov     _p,r1
40  mov     2(r1),r0
41  add     $2,2(r1)
42  add     $4,_p
43  mov     $7,(r0)
44  L1:jmp  cret
45  .globl
46  .data
```

Fig. 5.18. Example of the use of pointers and structures.

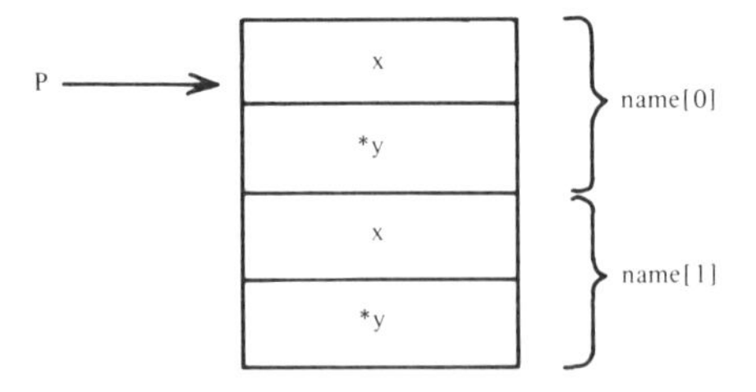

Fig. 5.19. Structure of variables in Fig. 5.18.

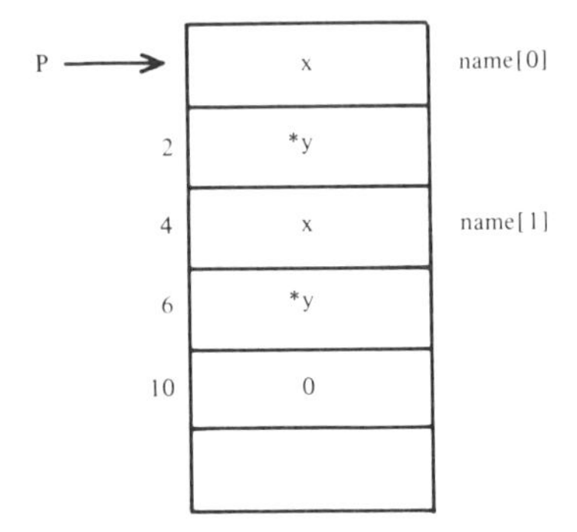

Fig. 5.20. p = name;.

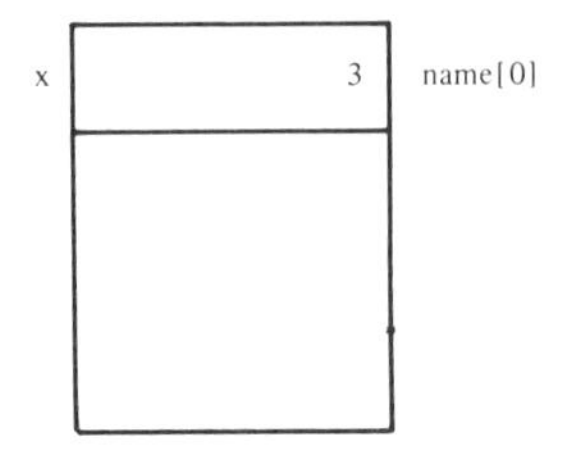

Fig. 5.21. p -> x = 3;.

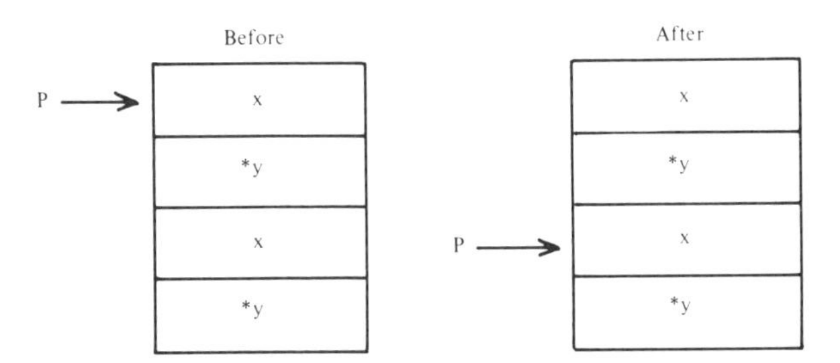

Fig. 5.22. Incrementing the pointer in the structure of Fig. 5.18.

Lines 12 and 31, 32

```
*p -> y = 4;  (C)
mov_p,r0      (as)
mov $4, *2(r0)
```

This instruction says to place 4 in the location pointed to by y. The -> binds more tightly than the * to the p. r0 first receives the address of p, then 4 is placed in the address r0 + 2. *2(r0) simply says use r0 + 2 as a pointer.

Lines 13 and 33, 34

```
*p++->y = 5;
mov   _p, r0
mov   $5, *2(r0)
add   $4,_p
```

This operation is the same as the previous example, except p is incremented after use. Thus, 4 bytes (= 2 words) are added to p to skip from name[0] to name[1] (see Fig. 5.22).

Lines 14 and 36–38: Exactly as before, note that the parentheses have the same effect.

Lines 15 and 37–43:

```
*(p++->y)++ = 7;
mov_p,r1        r1 has address of p
mov 2(r1), r0   *y is put in r0
add $2, 2(r1)   two is added to pointer
add $4,_p       p pointer is incremented for next structure
mov $7, (r0)    7 is put into where y is pointing
```

This is a complex statement. First, move the address of p to r1. Then place contents of r1 + 2 in r0 (this is *y). Next, add 2 to r1 + 2 (increments y). Adding 4 to_p skips to next structure and finally 7 is moved into the address pointed to by *y.

After some practice, the user of structures will find them to be a useful way to organize data. It is particularly useful in organizing blocks of repeated data. For example, if 10 variables are to be extracted from an experiment that is to be repeated 100 times, the 10 variables can be defined in one structure and a named array of 100 repetitions defined to have the form of the structure. This sort of definition keeps one from getting confused with multiple-dimensional arrays.

TABLE 5.3

COMMONLY USED SYSTEM CALLS

1.	chdir (directory name)	Change working directory
2.	close (file ID)	Close a file
3.	creat (name, mode)	Create a file
4.	exit (status)	Terminate a process
5.	fork ( )	Spawn a new process
6.	kill (pid, sig)	Send a signal to a process pid = process id, sig = signal
7.	mount (special, name, rw flag)	Mount a file system
8.	nice (priority)	Run a job with a priority
9.	open (name, access)	Open a file for reading or writing
10.	pipe (&2 word array)	Create an interprocess channel
11.	read (file descriptor, buffer address, #bytes)	Read from a file
12.	sleep (seconds)	Stop execution for "seconds"
13.	stty (fd, &3 word array)	Set tty parameters
14.	sync ( )	Update super block
15.	time (&2 word array)	Get time and date
16.	unmount	Dismount a file system
17.	wait (&status)	Wait for process to terminate
18.	write (fd, buffer address, count)	Write on a file

5.17 SYSTEM CALLS IN 'C'

Calls to the operating system in 'C' are carried out in much the same way as outlined for 'as' in Chapter 4. All calls are functions as shown in examples in Table 5.3. Calls frequently duplicate commands available in Section I of the programmers manual [e.g., chdir( ), stty( ), time( ), nice( ), etc.] Thus, one can conduct a series of operations in a program as if they were being conducted at the terminal keyboard. For example, the use of creat( ), write( ), and close( ) can be easily demonstrated. Suppose we wish to create a file containing the results of the integer square root program shown earlier. As indicated earlier, this operation is rather trivially conducted using the redirection facilities in the command line of the UNIX system. It can also be accomplished using the set of system calls. For example, in Fig. 5.7 add the following statements after the final printf command:

```
resultp = "result";
if ((output = creat (resultp, 0666)) <=0){
     printf ("creat failed\n");
     exit ( );
```

```
        }
        if (write(output,&count,2)!=2){
            printf (''write failed\n'');
            exit ( );
        }
        close (output);
```

```
main (argc, argv)
int     argc;
char    *argv[];
{
    int     output;
    char    *resultp;
    int     j,
            n,
            odd,
            count;

    n = j = atoi (argv[1]);
    count = 0;
    odd = 1;

    for (odd = 1; (n = n - odd) >= 0; odd = odd + 2) {
/*      printf("n-odd= %d and odd= %d\n",n-odd,odd );*/
        count++;
    }
    printf ("the integer root of %d is %d \n", j, count);
    resultp = "result";
    if ((output = creat (resultp, 0666)) <= 0) {
        printf ("creat failed\n");
        exit ();
    }
    if (write (output, &count, 2) != 2) {
        printf ("write failed\n");
        exit ();
    }
    close (output);
}
atoi (s)
char    s[]; {
    int     i,
            n;
    n = 0;
    for (i = 0; s[i] >= '0' && s[i] <= '9'; i++)
        n = 10 * n + s[i] - '0';
    return (n);
}
```

Fig. 5.23. Modified version of the integer square root program in Fig. 5.7 that allows writing the result to a file.

and at the beginning add the definitions:

```
char *resultp;
int  output;
```

The final program will now appear as shown in Fig. 5.23. In brief, a file name "result" is pointed to by resultp and a file is created with the name result. Mode 666 [permission to read/write—see chmod(I)] is specified. "creat( )" returns a file descriptor integer that is assigned to output. If creat( ) fails, a −1 is returned and the "if" statement causes the error message to be printed. Next, the 2-byte answer (count) is written out to result. The answer is in count; the address of count (&count) is passed to the write system call. If two bytes are not written, an error message is printed. Finally, the file is closed. Note carefully the parentheses construction in the creat call. To properly assign output to a file ID, parentheses must enclose the variable output and the system call. One can test the revised program by

```
% cc root.c
% a.out 64
```

The integer root of 64 is 8

```
% od result
000000  000010
000002
```

Thus the answer $10_8 = 8_{10}$ is the value stored in result.

The above example does not explain in detail the use of system calls, it only gives a flavor. Details of each call are given in the Programmer's Manual and there is an entire chapter devoted to the subject in "The C Programming Language" (Kernighan and Ritchie, Chapter 8). The interested reader is referred to these sources.

EXERCISES

1. Study the communication facilities of the UNIX system. One can write messages from one user to another or mail messages. What is the difference? See if you and another person can effectively communicate.
2. First, create a file using 'teco'. Then create the same file using the UNIX system editor 'ed'. From reading the manuals, describe the differences between teco and ed. Are there advantages and disadvantages to each?

3. Using the 'chdir' command, explore the UNIX system tree. Examine the contents of the directories starting at "root" or "/" on your system. How does your system differ from the system shown in Fig. 5.2?
4. If you are in your account and a friend wants a copy of a program you have, how would you transfer the program to your friend's account? Specify the steps you would take.
5. Examine the libraries in /lib. When would you want to use each of the libraries? Describe the contents of each library (use ar command) and indicate how to access these files in your programs. What type of files are they?
6. The "root" example in Section 5.10 produces only integer square roots. Can you use the same algorithm to obtain a more accurate result? Also, how could you obtain more accuracy without even changing the program?
7. The code generated by the pointer technique and the array technique (both optimized and not optimized) should be examined. Figures 5.24–5.27 show the results of producing four different .s files from

   (1) The array version of the add program, Section 5.13

      (a) optimized
      (b) nonoptimized

   (2) The pointer version of the add program

      (a) optimized
      (b) nonoptimized

   Discuss the differences in the code generated by the compiler in the four cases. Estimate differences in running time and calculate the size differences. Produce your results in the form of a table as shown in Table 5.4.
8. If the program for adding using pointers in Section 5.13 had been written using a structure for k[ ], would the results found in Exercise 7 have been any different?
9. Test out the example given in Section 5.17. Can you show that a file is created? Next, write a program to read the result file and print the result value to the standard output.

```
.data
.globl   _k
_k:
1
2
3
4
5
.text
.globl   _main
_main:
~~main:
jsr      r5,csv
~i=177770
~sum=177766
sub      $4,sp
clr      -10(r5)
L20001:mov       -10(r5),r0
asl      r0
mov      _k(r0),r0
add      -12(r5),r0
mov      r0,-12(r5)
inc      -10(r5)
cmp      $5,-10(r5)
jgt      L20001
mov      r0,(sp)
mov      $L5,-(sp)
jsr      pc,*$_printf
tst      (sp)+
jmp      cret
.globl
.data
L5:.byte 124,150,145,40,163,165,
155,40,151,163,40,45,144,40,12,0
```

Fig. 5.24. The optimized version of the program to add five numbers using an array k[ ].

```
.data
.globl   _k
_k:
1
2
3
4
5
.text
.globl   _main
_main:
~~main:
jsr      r5,csv
~i=177770
~sum=177766
sub      $4,sp
clr      -10(r5)
L2:cmp   $5,-10(r5)
jle      L3
mov      -10(r5),r0
asl      r0
mov      _k(r0),r0
add      -12(r5),r0
mov      r0,-12(r5)
L4:inc   -10(r5)
jbr      L2
L3:mov   -12(r5),(sp)
mov      $L5,-(sp)
jsr      pc,*$_printf
tst      (sp)+
L1:jmp   cret
.globl
.data
L5:.byte 124,150,145,40,163,165,
155,40,151,163,40,45,144,40,12,0
```

Fig. 5.25. The nonoptimized version of the program to add five numbers using an array k[ ].

```
.data
.globl   _k
_k:
1
2
3
4
5
.text
.globl   _main
_main:
~~main:
jsr      r5,csv
~i=177766
~p=177770
~sum=177764
sub      $6,sp
mov      $_k,-10(r5)
clr      -12(r5)
L20001:mov      *-10(r5),r0
add      -14(r5),r0
mov      r0,-14(r5)
add      $2,-10(r5)
inc      -12(r5)
cmp      $5,-12(r5)
jgt      L20001
mov      r0,(sp)
mov      $L5,-(sp)
jsr      pc,*$_printf
tst      (sp)+
jmp      cret
.globl
.data
L5:.byte 124,150,145,40,163,165,
155,40,151,163,45,144,12,0
```

Fig. 5.26. The optimized version of the program to add five numbers using pointers.

```
.data
.globl   _k
_k:
1
2
3
4
5
.text
.globl   _main
_main:
~~main:
jsr      r5,csv
~i=177766
~p=177770
~sum=177764
sub      $6,sp
mov      $_k,-10(r5)
clr      -12(r5)
L2:cmp   $5,-12(r5)
jle      L3
mov      *-10(r5),r0
add      -14(r5),r0
mov      r0,-14(r5)
add      $2,-10(r5)
L4:inc   -12(r5)
jbr      L2
L3:mov   -14(r5),(sp)
mov      $L5,-(sp)
jsr      pc,*$_printf
tst      (sp)+
L1:jmp   cret
.globl
.data
L5:.byte 124,150,145,40,163,165,
155,40,151,163,45,144,12,0
```

Fig. 5.27. The nonoptimized version of the program to add five numbers using pointers.

TABLE 5.4

PRESENTATION OF RESULTS FORMAT FOR EXERCISE 7

	Array with optimization	Array without optimization	Pointer with optimization	Pointer without optimization
Execution time				
Memory size required (bytes)				

REFERENCES

(1978) UNIX time sharing system. *Bell System Tech. J.*, **57**, No. 6, Part 2, pp. 1897–2312.

Kernighan, B. W., and Richie, D. M. (1978). "The C Programming Language." Prentice Hall, New Jersey.

Richie, D. M., and Thompson, K. (1974). The UNIX time-sharing system. *Comm. ACM* **17**, No. 7, 365–375.

6

I/O Fundamentals

6.1 INTRODUCTION

In a laboratory environment, fundamental uses of small computer systems are (1) acquiring data, (2) manipulating data or analyzing the data in some way, and (3) subsequently returning information to the external environment or experimental situation. Thus all laboratory computer users must be able to deal with I/O (input/output) manipulations. Users are commonly insulated from the intricacies of general I/O programming by operating systems that provide software interfaces (often called "handlers" because they "handle" the physical hardware devices) to common devices such as tape drives, disks, and various serial I/O devices (e.g., terminals). While such insularity is an attractive concept, many users often wish to use devices on their systems that do not have handlers already written. Persons inexperienced with operating system concepts are typically unprepared to deal with attaching hardware to a computer system that runs with an operating system that will not support that hardware. Thus there is a real need to know how to write device handlers or to communicate directly with specific devices from application programs. Typical needs are, for example, supporting an analog-to-digital converter, digital-to-analog converter, parallel digital I/O, or perhaps an unusual disk drive. Interfacing nonstandard peripheral devices to a program or to an operating system can be a simple or complex undertaking depending on the operating system. It is possible to access I/O devices directly from programs without an intervening system operations on most small computers found in laboratory environments. In contrast, in larger time-shared or multitasking environments (e.g., > 32Kw PDP-11s), an operating system interface (driver or handler) is typically written to facilitate generalized communication with the device. Large systems are, however, more frequently used for general time sharing than for laboratory data acquisition. Since most laboratory computing systems will allow direct access of peripherals, this text will have a major focus on direct peripheral communication rather than communication via device handlers.

All subsequent discussions will center around the PDP-11 and UNIX system although there will be a brief description of PDP-8 systems and the RT-11 operating system.

The two DEC computers, the PDP-8 and PDP-11, differ philosophically in the treatment of I/O. In the PDP-8, each I/O operation is assigned a unique 12-bit code. For example, the teleprinter and keyboard reader are each assigned six codes. Table 6.1 shows the codes for keyboard and printer operations. All other devices on the PDP-8 are assigned a similar set of codes. Note that codes begin with a 6 in the most significant octal digit of the 12-bit word to denote an I/O operation. Of the keyboard/teleprinter instructions, the KSF, KRB, TSF, and TLS symbols are most

TABLE 6.1

KEYBOARD AND PRINTER I/O CODES FOR THE PDP-8

Mnemonic symbol	Octal code	Operation
	Keyboard/reader (KB)	
KCF	6030	Clear KB flag
KSF	6031	Skip if KB flag set
KCC	6032	Clear KB flag and AC
KRS	6034	Read KB status
KIE	6035	Set/clear interrupt enable
KRB	6036	Read KB buffer
	Teleprinter (TP)	
TFL	6040	Set TP flag
TSF	6041	Skip on TP flag
TCF	6042	Clear TP flag
TPC	6044	Load on printer flag
TSK	6045	Skip on printer flag
TLS	6046	Load teleprinter and print

frequently used to read from the keyboard and write to the teleprinter. A typical PDP-8 operation sequence is

```
CLA          ; clear accumulator
TAD ("A      ; add ASCII value for the letter A
TSF          ; test to see if teleprinter is busy
JMP .-1      ; if busy, jump back one
TLS          ; if not busy print the A
```

The TSF instruction tests a flag (1 bit) that indicates whether the printer is ready to accept a character. If it is ready, the next instruction is jumped; if not ready, the next instruction is executed. Thus, the TSF, JMP .−1 statement pair is a simple wait loop. The TLS instruction simply prints the contents of the accumulator on the teleprinter.

In contrast, the PDP-11 uses memory locations for all I/O. All device registers on the 11 bear some resemblance to each other. More complex devices have more registers and simpler devices have fewer registers. On the PDP-11, typical registers used for many devices are a control register and a buffer register for data.

Each pair or set of device registers occupies sequential word locations in the last 4K words of memory. In nonmemory-managed systems, these locations are between 28K and 32K and in larger than 32K word PDP-11s with memory management (e.g., 11/34, 11/60, 11/23), the registers are addressed between 124K and 128K. Table 3.2 in Chapter 3 shows the locations of various I/O device registers. Most devices have a pair of allocated registers configured as shown in Fig. 6.1. The first register is the csr or control/status register, while the second register is the data register. Logically, if a device has, for example, multiple inputs or outputs, more registers will be required. For complex devices such as disk drives, up to 20 or 30 registers are frequently employed. Device registers are located in contiguous word locations and different devices may be located anywhere in the reserved 4K memory space.

A common bit assignment is shown in Fig. 6.1. Putting a one in bit zero of the control/status register (CSR) will start the device, while a one

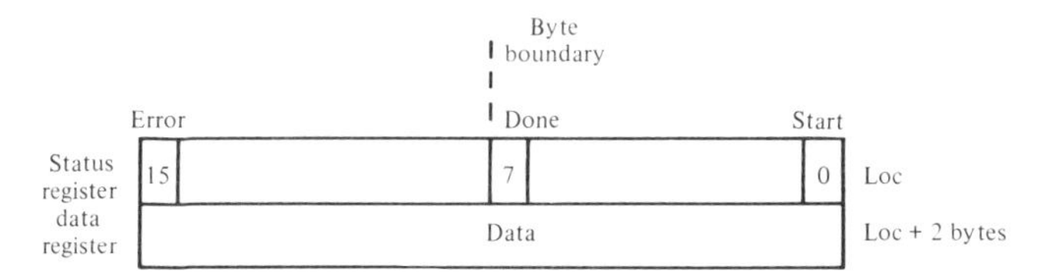

Fig. 6.1. Typical status and data register configuration.

appearing in bit 7 or 15 of the CSR will indicate that the device has completed an operation (bit 7) or that an error occurred (bit 15).

The attractiveness of the I/O methodology on the PDP-11 is that devices may be accessed in the same way as memory. Therefore, it is easy to use the control register to start a device or to test whether the device is ready, produced an error, etc. For example, one can test bit 7 (the done bit) by

```
loop:   tstb (r4)
        bpl loop
```

where r4 contains the address of the register of the device being tested.

Alternatively, the address can be specified directly in the tstb instruction, e.g.,

```
tstb *$0177560
```

or even

```
tstb 0177560
```

In the first case, the csr location (0177560) will be stored immediately following the instruction. In the second case, an offset from the PC will be generated to the location. A similar test statement in 'C' is

```
while (! (DEVREG->status & DONE));
```

The effect of the while loop is to wait until the contents of the loop become 0 to continue. DEVREG->status refers to the member of the structure:

```
struct { int status, data;};
```

where we have defined DEVREG by

```
#define DEVREG xxxxxx
```

and xxxxxx is the byte address of the device in question preceded by a 0 to indicate that the location is specified in octal. DONE is defined as the done flag (bit) by

```
#define DONE 0200
```

The contents of the status register and the 0200 are 'anded' to ensure that only the done bit is examined. Until bit 7 becomes set, the while argument !(. . .) is nonzero and the statement executes over and over. When bit 7 becomes 1, !1 = 0 is produced and the program proceeds. After a device that acquires data (e.g., keyboard, A/D) is started and completes its operation, the value acquired is available in the data register.

To move data from the data register to another location, the 'as' statement

```
mov *$DEVLOC+2, r4
```

will move the data to r4 if DEVLOC is the address of the control status word and the data word is in the next location. In 'C', the method of saving data is also straightforward. For example, after waiting for the device flag

```
while (!(DEVREG->status & DONE ));
*xpoint++ = DEVREG->data;
```

will store a data value wherever xpoint is pointing to and increment the pointer after using it.

Suppose we have allocated space for an array

```
int x[500];
```

and subsequently in the program set

```
xpoint = x;
```

xpoint will contain the address of x[0]. Note that xpoint must be declared also, for example, as

```
int *xpoint;
```

meaning that xpoint points to integers.

6.2 TERMINAL I/O

The most popular form of I/O for computer communication with users is via serial communication lines. As shown in Fig. 2.5 in Chapter 2, eight bits are usually sent serially for each character plus a start and stop bit. Of course, different length codes will use different numbers of bits. The most common code, however, is the 7-bit plus parity ASCII code. In the idle state, the line is set to a 1 and goes to a zero before sequentially sending the eight bits. A parity bit may be sent followed by a stop pulse and a return to an idle state. To receive serial information, data are fed into a shift register and from there fed in parallel to the computer. A done bit (or flag) is set to indicate that all bits have been shifted into the shift register. Figure 6.2 displays a conceptual block diagram of how data are transmitted to the computer. To send information from the computer to a serial device or line the reverse process is conceptually implemented as shown

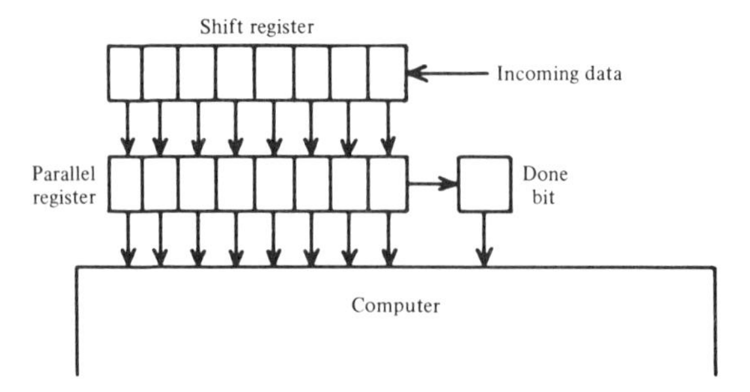

Fig. 6.2. Serial asynchronous receiver. In this diagram, the shift register shifts the incoming data left into the register.

in Fig. 6.3. After bits are loaded into a shift register, either directly or via a buffer, the bits are shifted out into the data line. When all zeros remain, their complement 'anded' produces a flag to indicate that shifting is complete.

Since there are entire texts written on the subject of data communication hardware, this text does not describe in any detail information about hardware for data communication. An excellent book that contains extensive information about hardware and communications protocols is entitled "Technical Aspects of Data Communication" (John E. McNamara, 1977; Digital Equipment Corporation). Next, specific ways of accessing data on serial communication lines connected to the PDP-11 series of computers will be examined.

Figure 6.4 shows the configuration of the device registers for the terminal printer and keyboard. The addresses shown are for a 32K word ma-

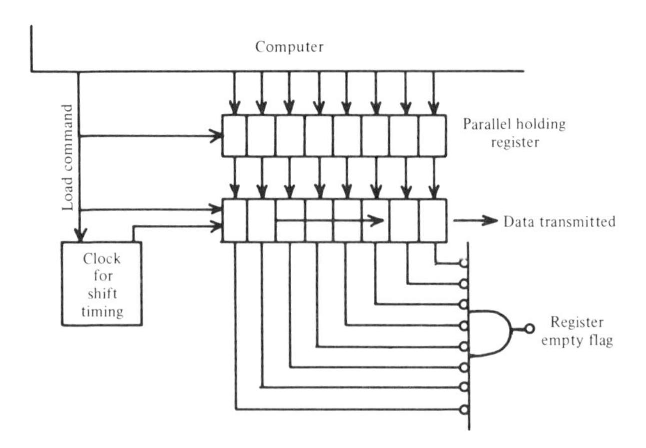

Fig. 6.3. Serial data transmission.

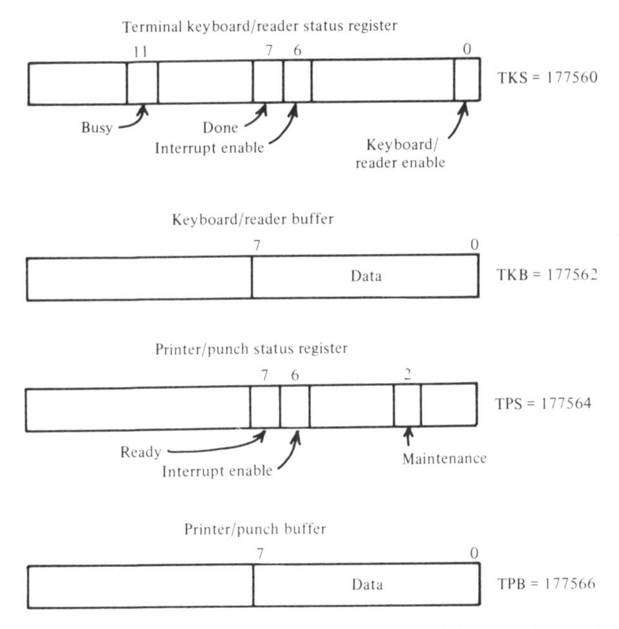

Fig. 6.4. Device register configuration for the console serial device located from 0177560 to 0177566.

chine with the device registers located between 28K and 32K. In a 128K word machine, these registers are located between 124K–128K. For example, address 0177560 becomes 0777560 in 128K address space. Note that the only difference in the address is the addition of the leftmost bits. This particular set of registers is for the first serial port, typically used for the console device. The console is the terminal from which the system is initialized and error messages are printed. As other individual lines are added, they will be assigned different addresses. Although a number of individual serial lines can be added in this way, it is often simpler to add a set of multiplexed lines (e.g., 8 or 16) that will allow access of all lines via a single set of registers (e.g., DZ-11 or DH-11 serial asynchronous multiplexers).

6.3 TERMINAL PROGRAMMING

Device registers are accessed as memory locations. This facilitates manipulation using standard instructions; e.g.,

```
mov $07, tpb
```

will move the value 7 to the teleprinter buffer. By assigning mnemonics to the device locations, they can be easily referred to using standard "as" instructions.

It is convenient in both 'C' and 'as' to define symbolic names to represent the device register locations. Common assignments in 'as' are

```
tks = 0177560    keyboard status
tkb = 0177562    keyboard buffer
tps = 0177564    printer status
tpb = 0177566    printer buffer
```

or

```
tks = 0177560
tkb = tks + 2
tps = tks + 4
tpb = tks + 6
```

As shown in Fig. 6.4, bit assignments are similar to the model given previously for all I/O devices. Consider a simple program in 'as' named echo.s that prints everything that is typed on the keyboard back on the teleprinter:

```
tks = 0177560
tkb = tks + 2
tps = tks + 4
tpb = tks + 6

 start:  bis $1, tks      ; enable keyboard
loopr:   tstb tks         ; is bit seven set?, if not
         bpl loopr        ; we branch to loopr when the
                          ; byte is positive
loopp:   tstb tps         ; ready to print? wait until
         bpl loopp        ; bit comes up
         movb tkb, tpb    ; input to output
         br start         ; go to start
```

On an 11/03 running RT-11, one can add a header to the beginning of this program to allow the compiled code to start at some specific location in memory (e.g., 1000). Using the transfer system listed in the Appendix and discussed in Chapter 5, the following steps will run echo.s.

```
% as – echo.s
% out-sav a.out   [fixes header]
% to03vt a.sav    [transfers file to 11/03 in name 03.DAT]
.R 03.DAT (on RT-11) (program runs)
```

The most interesting feature in the above example is the use of the tstb instruction, which allows testing a single byte. When bit seven is set (e.g., by the completion of the teleprinter), the result will be the setting of the N bit (negative) in the PSW (processor status word). One can check the keyboard buffer (tkb) for specific characters to control the flow of a program. For example,

```
        .
        .
        .
mov tkb, r0
bic $0177600, r0
cmpb r0, $'g
beq go
        .
        .
        .
```

Some terminals produce a parity bit in the most significant bit in the first byte in tkb. Thus it is prudent to clear this bit before doing a comparison between the ASCII value typed and an ASCII value stored in the program. Since tkb is read-only, the value in tkb should be moved to a register and the bits cleared prior to checking. This type of testing will work well on a simple serial card but will not work on a more complex multiplexed device such as a DH-11 or DZ-11 on which values are stored on a silo (think of a silo as a queue).

Suppose we design a program to start (any device; e.g., sampling data) when a control s (^s) is typed on the keyboard, stop (^z), display (^d), etc. As shown in Fig. 6.5, we sit in various loops forever until a ^s, ^z, ^r, ^c, or ^d is typed. At the outset, we wait until a ^s is typed; nothing else can cause the program to proceed. An alternative would be to include a check for a standard exit such as ^c in each loop. That is,

```
start:  mov tkb, r0
        bic $0177600, r0   / clear all bits except 0-6
        cmpb r0, $'03
        beq exiter
        cmpb r0, $'023
        beq sample
        br start
exiter: sys exit
```

In this way we cannot get hung in a mysterious loop with no perceivable exit. Another possibility is to type a message prior to beginning that

```
start:  mov tkb,r0              /clear all but the
        bic $0177600,r0         / least significant 7 bits
        cmpb r0,$023            / control s = 023
        beq sample              /wait in loop
        br start


sample:                         / start sampling


loop:   mov tkb,r0
        bic $0177600,r0
        cmpb r0,$032            / control z
        beq stop
        br loop

stop:                           / stop sampling

loop1:  mov tkb,r0
        bic $0177600,r0
        cmpb r0,$04             / control d
        beq display
        cmpb r0,$022            / control r
        beq reset
        cmpb r0,$03             / control c
        beq exitt


        br loop1

reset:                          / reset here

exitt:  sys exit
```

Fig. 6.5. Example design strategy for using control characters on the keyboard to control program flow.

says, for example,

"Type a carriage return to begin sampling:"

Normally, however, if an interrupt routine is used, or if an operating system is involved, standard exit commands such as ^c will be checked by the operating system as they are typed. This checking is one virtue the user receives while being insulated from devices by the operating system. However, if one is running a program without the services of an operating system, one must attend to such details. In an interrupt routine (see Chapter 8), each character can be checked against a list every time a character is typed.

Note in the code in Fig. 6.4 that the label "exitt" has an extra "t." This spelling rather than "exit" is necessary because the 'as' assembler has reserved "exit" for a system call. If you experience problems in assembly, labels that are the same as system call names will definitely cause problems. Beware!

6.4 TESTING DEVICE REGISTERS USING ODT ON AN 11/03

ODT (octal debugging technique) (see Chapter 2) makes it possible to easily test I/O devices. For the terminal, a simple example is

```
@ 177566/000000 07 RET
```

The bell (buzzer, tone) will ring because 07 corresponds to control G, the code to activate the bell. Also, other character codes can be typed in, e.g.,

```
@177566/0144 RET
```

will cause an 's' to appear on the screen.

6.5 'C' AND THE TERMINAL

The device registers are easily specified by use of a structure:

```
struct { int tks,
             tkb,
             tps,
             tpb;};
```

This structure defines a block of four integers with the names shown. Or, to allow byte access:

```
struct { char tks, filler1,
              tkb, filler2,
              tps, filler3,
              tpb, filler4;
       };
```

where "filler1" to "filler4" reference the high byte of each word and the teleprinter names, the low byte. Thus,

```
while(!(DEVLOC->tps));
```

for the byte structure will do the same job as

```
while(!(DEVLOC->tps & DONE));
```

for the word structure. For convenience of notation, we shall refer informally in this text to the word definitions above. The byte usage may, however, be somewhat faster (see Exercise 5).

In the UNIX* system, using memory-managed systems, one can only access the device registers via a device handler. That is, operations by a user that insert values (or read values) from locations in the upper 4K words of memory are detected and the user is denied access by the UNIX operating system. Thus, users must access device registers via either built-in system primitives or build primitives into the operating system. By primitive we mean a single simple operation. For example, the UNIX system provides the system calls or primitives "getc" and "putc" for the terminal. These routines are available either from 'C' or from 'as'. "getc", "putc", "getchar", and "putchar" are system calls that allow accessing characters. There are also routines for opening a file and using words rather than characters. For most users, the calls getchar( ) and putchar( ) are easiest to use. These calls read or write characters from or to the standard input or output (can access other terminals; recall the redirection facilities) until an end of file (EOF) = \0 is encountered. The value of the use of "getc" and "putc" is that different files can be accessed.

Some refer to the above types of calls as "hooks" into the operating system. Of course, other calls for other devices could be written, for example, to read an integer from the A/D, plot a point on a D/A, etc.

Various printing- and character-reading routines such as "printf" and "scanf" use "putc" and "getc" in their routines. The UNIX/'C' system user with no aspirations for using 'C' standalone on an 11/03 does not need to know how printf works. If, however, the user would like to run the routine on a small system, e.g., a PDP-11/03 or perhaps change the code (to condense it, for example), he will need to understand how it works. We shall present an example later in this chapter that describes "printf" modifications.

6.6 PDP-11/03, 'C', AND THE TERMINAL

If one prepares the simple 'C' program called test.c,

```
main( ) {printf ("This is a test\n");}
```

* UNIX is a trademark of Bell Laboratories.

assembles and runs it, the message shown will appear on one's terminal. For example,

```
%cc test.c
%a.out
This is a test
%
```

Yet, if one takes the very same a.out file, puts it into memory on the 11/03, the program will not run. Why? Because printf calls putchar( ), and putchar( ) is provided by the UNIX system to facilitate service requests on the operating system. Thus if we wish to run test.c on an 11/03, we must provide our own character-handling routines. Consider Fig. 6.6, a copy of the conventional printf routine reprinted from the 'C' reference manual. As shown, there are a number of different format possibilities including %d (decimal), %c (character), %f (floating), %s (string), %o (octal), which are tested in the program switch statement on line 15. We shall modify printf to obtain the scaled-down version called print.c shown in Fig. 6.7. This routine has only three options for printing, %d, %c, and %s. Two arguments are passed, a pointer to the format string (i.e., the characters contained within quotes), and the pointer to the argument pointers. An endless loop is entered in line 10,

for (;;) (almost the same as while (1))

and characters are put out with putchar (c) until a % is reached in the format line. Thus, if a %c is encountered, characters are accessed via the arg pointer (ap = &args;) and we call putchar(*ap++); i.e., we get the address of the arg list on the stack, get the character indirectly and increment the arg pointer ap. The same sort of operation occurs for %s except we have a double pointer. Strings are accessed via an address that points to the beginning of the string. Thus, ap.charpp refers to pointers to pointers, s is the pointer to the string, and c = *s++ accesses the individual characters.

The decimal print is somewhat different, however. After the integer is accessed by x=*ap++, printd(x) is called, which converts the integer value into a printable set of characters. "printd(x)" is a recursive routine (calls itself). Consider, for example, that the number x = 541 is passed to printd(n). The algorithm divides 541/10 = 54 = a. If the result (here 54 ! = 0) is ! = 0, it calls itself with the argument 54. On this iteration 54/10 = 5, etc. Each time the routine calls itself, the variables are pushed onto the stack. On the last iteration, the "if" statement is no longer !=0 (5/10 = 0) and the putchar statement is executed. That is,

5%10 + '0' = 5

```
 1  printf(fmt,args)
 2  char fmt[];
 3  {
 4          char *s;
 5          struct{char **charpp;};
 6          struct{double *doublep;};
 7          int *ap,x,c;
 8          ap = &args;             /*argument pointer */
 9          for(;;){
10                  while((c = *fmt++)!='%'){
11                          if(c == '\0')
12                                  return;
13                          putchar(c);
14                  }
15                  switch(c = *fmt++){
16                  /* decimal */
17                  case 'd':
18                          x = *ap++;
19                          if(x<0){
20                                  x = -x;
21                                  if(x<0){ /* is - infinity */
22                                          printf("-32768");
23                                          continue;
24                                  }
25                                          putchar('-');
26                                  }
27                                  printd(x);
28                                  continue;
29                          /* octal */
30                          case 'o':
31                                  printo(*ap++);
32                                  continue;
33                          /* float,double */
34                          case 'f':
35                          /* let ftoa do the real work */
```

Fig. 6.6. The printf function for printing decimal, octal, and floating numbers, characters, and character strings. Reprinted from the 'C' reference manual with permission of the Bell System.

is printed. 5 is the remainder of 5/10 and '0' = 60. The "%" symbol is used to produce the remainder of the division of two numbers. Thus 65 (ASCII) is produced, which when fed to a device register by putchar prints the character 5. Next, the stack is popped up, restoring the variables and

$$54\%10 \text{ is printed} = 4 + \text{`0'}.$$

Finally,

$$541\%10 \text{ is printed} = 1 + \text{`0'}.$$

Thus this is a very short printing algorithm for decimal numbers.

```
36                                  ftoa(*ap.doublep++);
37                                  continue;
38                          /* character */
39                          case 'c':
40                                  putchar(*ap++);
41                                  continue;
42                          /* string */
43                          case 's':
44                                  s = *ap.charpp++;
45                                  while(c = *s++)
46                                          putchar(c);
47                                  continue;
48                          }
49                          putchar(c);
50                  }
51          }
52          /*
53          *Print n in decimal; n must be non-negative
54          */
55          printd(n)
56          {
57                  int a;
58                  if(a=n/10)
59                          printd(a);
60                  putchar(n%10 + '0');
61          }
62          /*
63          *Print n in octal, with exactly 1 leading 0
64          */
65          printo(n)
66          {
67                  if(n)
68                          printo((n>>3)&017777);
69                  putchar((n&07)+'0');
70          }
```

Fig. 6.6. (continued)

To convert the print.c routine for use on the 11/03, all the putchar statements must be replaced. Each putchar is replaced by two statements:

```
while (!(KLPRINT->tps&KLDONE));
KLPRINT->tpb = c;
```

The capital letter names above are defined in Fig. 6.8. Note that it is convenient to maintain a file with definitions—in this case tty.h. In a main program that uses these definitions, they may be included by

```
#include "tty.h"
```

```
print(fmt,args)
        char fmt[];
{
        char *s;
        struct { char **charpp;};
        struct { double *doublep; };
        int *ap,x,c;

        ap = &args;
        for (;;) {
                while((c = *fmt++) != '%'){
                        if(c == '\0')
                                return;
                        putchar(c);
                }
                        switch (c = *fmt++){
                                case 'd':
                                        x = *ap++;
                                        printd(x);
                                        continue;
                                case 'c':

                                        putchar(*ap++);
                                        continue;

                                case 's':

                                        s = *ap.charpp++;
                                        while(c = *s++)
                                                putchar(c);
                                        continue;
                        }
                        putchar(c);
                }
}


printd(n){

        int a;
        if(a=n/10)printd(a);
        putchar(n%10 + '0');
}
```

Fig. 6.7. A modified version of Fig. 6.6 that permits printing only decimal numbers, characters, and character strings.

```
define KLINTV 060
define KLREAD 0177560
define KLPRINT 0177564
define KLPSW 0340
define KLDONE 0200
define INTON 0100

truct {
   int      tps,
            tpb;
;                             /* status and printer buffer */
truct {
   int      intvec,
            psw;
;                             /* interrupt vector and processor status word */
truct {
   int      tks,
            tkb;
;                             /* keyboard status and printer buffer */
```

Fig. 6.8. Definitions file. This file may be inserted in programs by using the statement #include ''tty.h'' in other programs that use the definitions in the file. For use with other example programs in this chapter, name this file tty.h. Note the requirement for inclusion of a # as the first character in code that uses the ''include'' statement.

if tty.h is in the same directory. Otherwise you must use path names, e.g.,

#include ''/usr/john/ttytest/tty.h.

If an include statement is used in your program, the first character in the first line of the program *must be* a #. The modified version of Fig. 6.7 is shown in Fig. 6.9. Consider Fig. 6.8 further. After the definitions, three structures are given: (1) for the teleprinter, (2) for the interrupt vector, and (3) for the keyboard. As discussed in Chapter 5, a structure provides a convenient way of organizing information. By defining

```
struc {int tps,
           tpb;};
```

a structural form of two words is set up. One can consider these integers to be a block of two words located anywhere in memory. When the structure members are referenced with respect to an address (e.g., KLPRINT), the structured block of definitions can be viewed as being placed at that location in memory with the first definition in the structure corresponding to the address. Hence KLPRINT−>tps refers to location 0177564, and KLPRINT−>tpb refers to location 0177566. The structure members (here tps and tpb) are simply offsets from the basic name (KLPRINT).

```
#
#include "/usr/john/ttytest/tty.h"
print (fmt, args)
char    fmt[];
{
    char   *s;
    struct {
        char  **charpp;
    };
    struct {
        double *doublep;
    };
    int    *ap,
            x,
            c;

    ap = &args;
    for (;;) {
        while ((c = *fmt++) != '%') {
            if (c == '\0')
                return;
            while (!(KLPRINT -> tps & KLDONE));
            KLPRINT -> tpb = c;
        }
        switch (c = *fmt++) {
            case 'd':
                x = *ap++;
                printd (x);
                continue;
```

Fig. 6.9. Modification of Fig. 6.7 to permit direct access of terminal registers for printing.

This scheme does not have to be used for accessing the device registers. In the program we could, for example, set

$$\text{tps} = 0177564;$$

and then test tps, i.e.,

$$\text{while } (!((*\text{tps})\&\text{KLDONE}));$$

6.7 TEST PROGRAMS

Figure 6.10 shows a simple test program in 'C' and the .s equivalent using the print function we described above. Note carefully the use of *print* instead of printf! Also note that we must have a \n and a \r at the end of a string. \n = 12 and \r = 15 must both be supplied (newline and return). When using UNIX or MINIUNIX system, the operating system supplies

```
                case 'c':

                    while (!(KLPRINT -> tps & KLDONE));
                    KLPRINT -> tpb = *ap++;
                    continue;

                case 's':

                    s = *ap.charpp++;
                    while (c = *s++) {
                        while (!(KLPRINT -> tps & KLDONE));
                        KLPRINT -> tpb = c;
                    }
                    continue;
            }
            while (!(KLPRINT -> tps & KLDONE));
            KLPRINT -> tpb = c;
        }
}


printd (n) {

        int     a;
        if (a = n / 10)
            printd (a);
        while (!(KLPRINT -> tps & KLDONE));
        KLPRINT -> tpb = (n % 10 + '0');
}
```

Fig. 6.9. (continued)

the other when you supply one. This test program can be run with the standard transfer system:

% cc −2 test.c print.c	[−2 adds header for RT-11]
% out-sav a.out	[produces a.sav]
% to03vt a.sav	
on 11/03:	
.R 03.DAT	

The program can also be transferred to the 11/03 using the transfer system for parallel lines. The loader "minild" may be used to prepare files to run at 060000 in memory, if the MINIUNIX operating system is used. The steps for preparing a program named junk.c to run on this system are

% cc −c print.c
% cc−c junk.c
% minild /lib/crt0.o junk.o print.o −1C
% transfer a.out

```
main () {
    print ("this is a test \n\r");
}


/ shown below is the .s version of the above program



.text
.globl  _main
_main:
~~main:
jsr     r5,csv
mov     $L2,(sp)
jsr     pc,*$_print
L1:jmp  cret
.globl
.data
L2:.byte 164,150,151,163,40,151,163,40,141,40,164,145,163,
164,40,12,15,0
```

Fig. 6.10. A simple 'C' program using the print function defined in Fig. 6.9. The ".s" code equivalent of the 'C' code is also shown.

The library routine crt0.o contains standard startup code information and −1C retrieves cret and csv from /lib/libC. "transfer" transfers the file to the 03, storing in memory beginning at 60,000, and then runs the program. Of course, the loader step could be avoided and cc used directly if a 060000 header is loaded at the beginning of the compiled file, e.g.,

```
% make header.as
  I .=.+60000
  *ex$$
% as header.as
% mv a.out header.o
% cc − c junk.c print.c − 1c
% 1d header.o /lib/crt0.o junk.o print.o −1c −1
```

Thus, in the load module, "header" creates a 060000 open area before junk.o gets loaded at 60,000.

Note carefully that these steps are not necessarily the ones to use on other systems, since each configuration is usually somewhat different. The explicit steps are included here for two reasons: (1) to give the concept definition in terms of actual instructions and procedures, and (2) to pro-

```
#

#include "tty.h"        /* the definitions for the tty registers */

main () {
    for (;;) {
        while (!(KLPRINT -> tps & KLDONE));
        KLPRINT -> tpb = 07;
    }
}


/ Shown below is the bell.s file produced by
/ using the command string "cc -S bells.c"
/ on the above program.


.text
.globl  _main
_main:
~~main:
jsr     r5,csv
L2:L4:bit       $200,*$-214
jne     L5
jbr     L4
L5:mov  $7,*$-212
jbr     L2
L3:L1:jmp       cret
.globl
.data
```

Fig. 6.11. "bells.c". A program to sound continuously the bell on the terminal.

vide a guide for those who undertake to implement the programs listed in the Appendix.

Figure 6.11 displays a second example that rings the bell on the terminal. A simple wait loop precedes stuffing tpb with a 07 (control G—rings bell). Shown below the 'C' program is the .s equivalent of the 'C' file. Note how easy it is to verify that '07' is actually put into the tpb.

EXERCISES

1. Many 11/03 systems have multiple serial ports connected, each with different device addresses. Write a program to print out information on a terminal connected to one port when information is typed on a terminal connected to another port.

(a) If you can perform the above communication, can you write from a file to any serial line?
(b) If you connect a serial line to a modem and write from a file to another computer connected via the serial line, you should be able to transfer data between computers. Is this as easy as it sounds? See if you can write a program to accomplish this task.

2. Write a small program called a "bells." The function of this program is to alert the programmer by repetitively beeping when a job is complete. For example, after typing

% cc job.c

the programmer may have to wait for a long time. If the user types

bells

after the cc command line, then "bells" will be executed when cc job.c is completed. Specify parameters for "bells" and try it out.
3. Text in the English language has large numbers of blanks and vowels repeated. A program exists in the UNIX system for packing files such that the most frequently used characters are given the shortest codes. This program is useful for compressing files to allow less disk space to be used. For one of your files:

% pack yourfile.c
produces yourfile.c.p

Do an ls − 1 on both files. What is the file size in bytes? Why? Is this a good way to store data? (Note, you can cat packed files with % pcat fname.)
4. Figure 6.12 shows an 'as' program for typing 'hello'.

```
        tps = 177564
        tpb = 177566

hello:  mov $list, r3
loop:   tstb tps
        bpl loop
        cmpb *r3,$'\0
        beq done
        movb *r3,tpb
        inc r3
        br loop
done:   sys exit
list:   <hello\n\r>
```

Fig. 6.12. An 'as' program for typing 'hello' on the console terminal.

(a) Explain how this program operates.
(b) Run the code on an 11/03. Are any modifications necessary to make it run correctly?

5. Is the assertion about access speed in Section 6.5 using the byte structure true? Can you prove whether it is correct or incorrect?
6. What are the differences between the following two statements?

```
for (;;)    {. . .}
while (1) {      }
```

REFERENCES

McNamara, John E. (1977). "Technical Aspects of Data Communication." Digital Equipment Corporation, Maynard, Massachusetts.

7

Laboratory I/O—A/D, D/A, Clocks

7.1 INTRODUCTION

When a computer is used to analyze signals acquired from the external world, the signals must be first digitized (i.e., converted from analog signals to digital values). An analog-to-digital converter (A/D) is used for this purpose. A digital-to-analog converter (D/A) is employed for converting digital values into analog signals. Thus for these two devices the interfaces between the computer and the exterior world appear as shown in Fig. 7.1.

Several analog input channels (e.g., 8, 16, 32, 64) can be fed into an analog multiplexer (MUX), which presents the A/D converter with one analog input to the sample at a time. In a typical system, multiple channels cannot be sampled simultaneously. In practice, the MUX is rapidly switched between channels to achieve near-simultaneity. Sampling typically occurs in a few microseconds (e.g., 20) and the converted value is available to be used in a program.

Commonly found input voltage levels to an A/D are ± 1, ± 5, or ± 10 V. Most available A/D cards allow the user to choose an input voltage level by selecting a specific jumper on the card. Similarly, D/A outputs can be adjusted by changing jumpers to allow different output voltage levels. There are three basic ways that one can add an A/D to a minicomputer system:

1. Use a single-packaged hybrid system that can be interfaced to a parallel input on the computer.

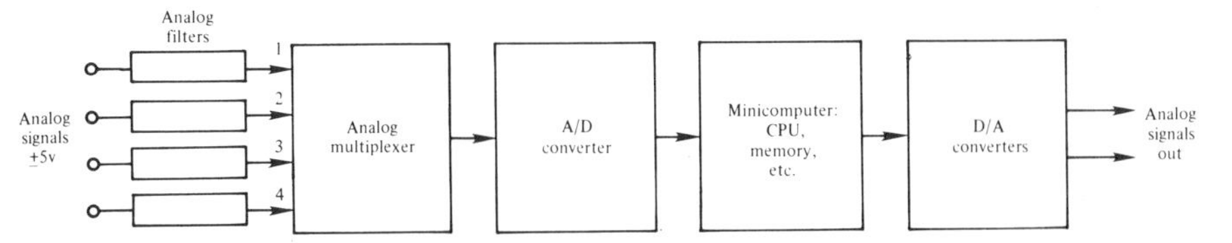

Fig. 7.1. Connection of A/D and D/A converters to a computer.

2. Purchase a complete card that plugs into a specific type of computer bus (e.g., the PDP-11) that can be used in the same way as other I/O devices. The purchaser of such a system should be aware of the specific bus configuration of his computer. For example, for the PDP-11, A/D cards can be purchased that will plug directly into the PDP-11/03-type bus (the "Q" bus) or into the Unibus on larger PDP-11s (e.g., the 11/34, 11/45, etc.). A card designed for one bus will not work for the other bus without a special converter.
3. Fabricate an A/D from a set of chips that can be connected together on a foundation module (see discussion in Chapter 3).

Choice 2 is the most straightforward. However, many commercial systems provide eight or more channels as a standard feature. In the interest of cost saving for a single-channel application, a custom arrangement may be more suitable. However, a common experience is that the effort of fabricating and debugging a home-built A/D is considerably greater than the differential cost between buying the parts and buying a complete tested unit. This chapter will focus on use of commercially available systems.

7.2 HOW D/As AND A/Ds WORK

Figure 7.2 shows a simple schematic diagram of a D/A converter. D/A converters are commonly available on chips or on cards that plug directly into minicomputers or are perhaps part of another system (e.g., an A/D card). One can conceptualize the operation of a D/A as shown in Fig. 7.3. A simple summer using operational amplifiers approximates the operation of a D/A. As shown, each bit contributes

$$\frac{R_F}{R} \times V_{in} \text{ volts} \times 2^{\text{bit position}}$$

Thus the output voltage will be proportional to the binary input data.

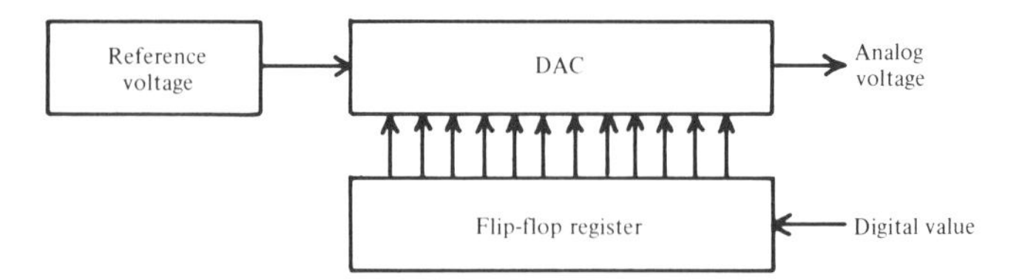

Fig. 7.2. Representation of a D/A converter.

Figure 7.4 shows a simple A/D converter fabricated from three comparators. The table accompanying this figure shows which comparators are on or off when the input voltage levels are in four different voltage ranges. Three comparators allow 3 bits to be represented. For an 8-bit converter, eight comparators would be needed. This simultaneous conversion method has the advantage of being fast but also expensive when high precision is needed.

Figure 7.5 displays three versions of the so-called feedback methods in which an input signal is compared with the output of a D/A converter. The general form of a feedback system is shown at the top of the figure. In one method, the counter method, a counter is allowed to count until the analog equivalent of the counter value equals the input signal level as shown in Fig. 7.5a. In the continuous method, the same system is used, except the counter can count either up or down, an obviously more efficient method. Still more efficient is the successive approximation method, in which the input range is successively divided in half, asking the question: "Is the voltage in the upper or lower half of the window?" When the voltage is in

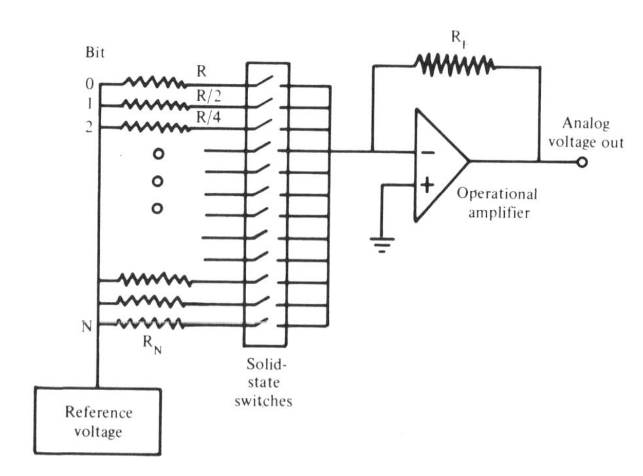

Fig. 7.3. A simple D/A converter using an operational amplifier.

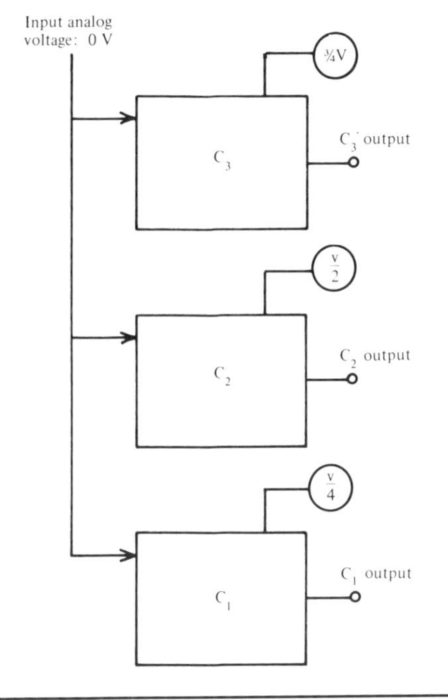

C_1	C_2	C_3	Input voltage
off	off	off	0–v/4
on	off	off	v/4–v/2
on	on	off	v/2–3v/4
on	on	on	3v/4–v

Fig. 7.4. The simultaneous method of A/D conversion.

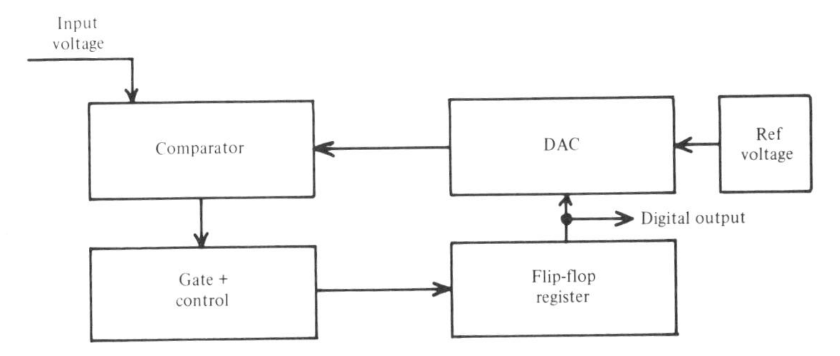

Fig. 7.5. Methods for A/D conversion. A general diagram applicable to all feedback systems is shown above. The next two pages show (a) counter method, (b) continuous method, and (c) successive approximation.

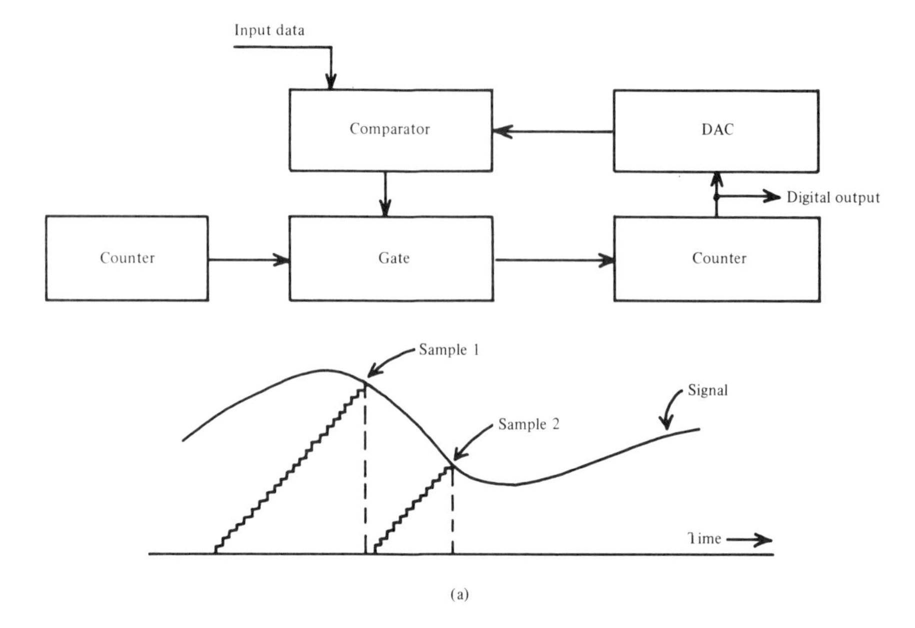

(a)

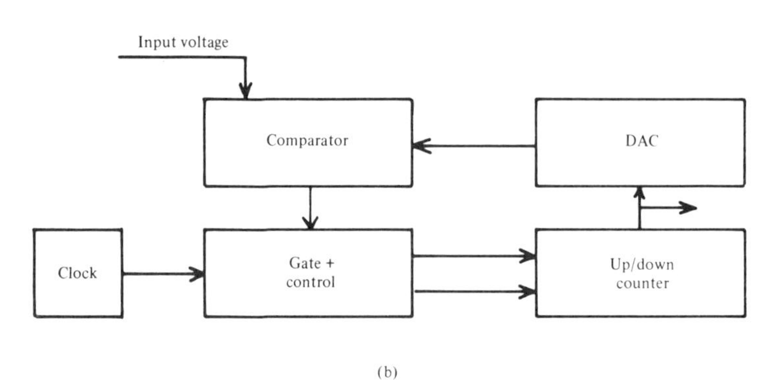

(b)

Fig. 7.5. (continued)

the top half of the window, a one is assigned and when the voltage is in the lower half, a zero. In the example in Fig. 7.5c, three successive approximations are required to determine that the voltage "x" is represented as a 010 in this eight-level window. This technique provides rapid convergence to a final answer.

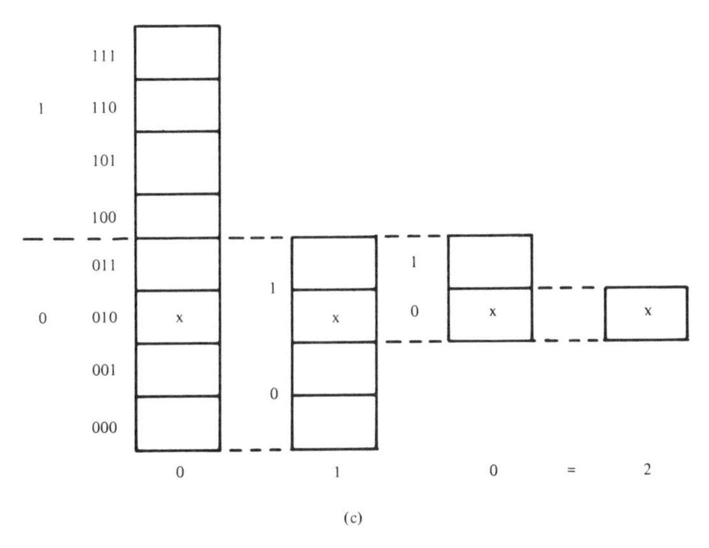

(c)

Fig. 7.5 (continued)

7.3 SAMPLING HARDWARE: REGISTERS AND VECTORS

Figure 7.6 displays the specifications for a common A/D and /DA card manufactured for the PDP-11/03, the model 1030 made by ADAC Corporation. This card is similar to the AR-11 card manufactured by DEC (see the DEC Peripherals Handbook). The model 1030 will be described; other cards have conceptually similar characteristics.

When an A/D card is obtained for the LSI-11, jumpers on the board must be selected for the address of the status and data registers and for the D/A locations. A common error is to plug a card in without checking the jumpers on the card. On the model 1030 the most important jumpers to be set are

1. Status register: 0176770.
2. Data register: 0176772.
3. D/A 1: 0176760.
4. D/A 2: 0176762.
5. Interrupt vector: 0130 address.
   0132 PSW.
6. Output voltage range.
7. Input voltage range.
8. Type of input (single ended, differential amplification in MUX).

ANALOG HIGH LEVEL DATA ACQUISITION SYSTEMS

ADAC MODEL	COMPATIBLE WITH
1030 Series	ADAC 1000 System DEC LSI-11 PDP 11/03

FEATURES

- **High-speed, 12 bit A/D converter**
- **High-speed sample & hold**
- **Up to 64 channels of MUX**
- **35/100 Khz throughput rate**
- **Plugs directly into computer mainframe**
- **Single-ended or differential inputs**
- **Programmable gain option with auto zero**
- **2-channel, 12 bit D/A converters**
- **Progam interrupt interface**
- **Includes cabling to rear panel**
- **Steel-cased modules for minimum RFI, EMI interference**

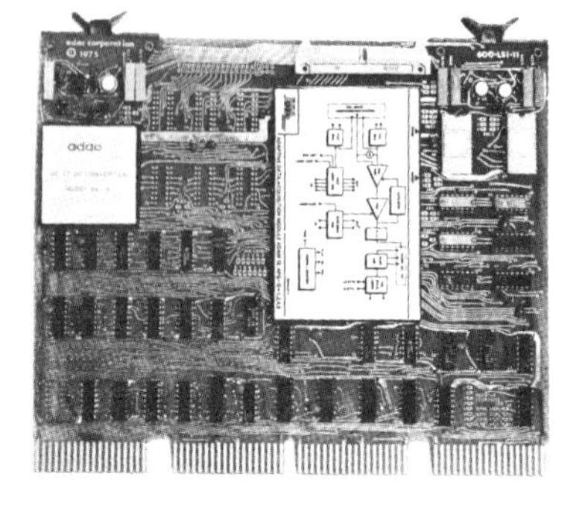

GENERAL DESCRIPTION

The ADAC 1030 Series of data acquistion systems can contain 16 to 64 channels of multiplexed 12 bit analog to digital converter inputs built on a quad size (8½" x 10") DEC style printed circuit card. It is designed as a companion peripheral for the DEC LSI-11 series of low cost 16 bit micro-computers or ADAC 1000 System to provide an economical connection to the real world.

The multiplexers of the 1030 Series can be connected either single-ended or differentially, and can operate either in the sequential mode or random access mode. A high speed sample and hold amplifier is employed as a buffer between the multiplexer and the ADC, thereby minimizing aliasing errors and providing for the digitizing of higher bandwidth signals.

The 12 bit analog to digital converter utilizes temperature compensated current switches and reference voltages to provide stable operation over a wide temperature variation. Full advantage is taken of the latest in LSI circuitry to minimize the part count of the ADC.

Four jumper selectable ranges are provided to allow tailoring of the system to specific application requirements. The 35/100 Khz system throughput rate allows rapid accumulation of input data from a large number of sources.

A software programmable gain amplifier can be provided to allow the computer to select one of four gain settings for any of the input channels over a 10 to 1 range. Included with the amplifier is an automatic zeroing feature that prevents offset drifts even at the highest gain. Provision is also made for two 12 bit D/A converters to provide analog output capability.

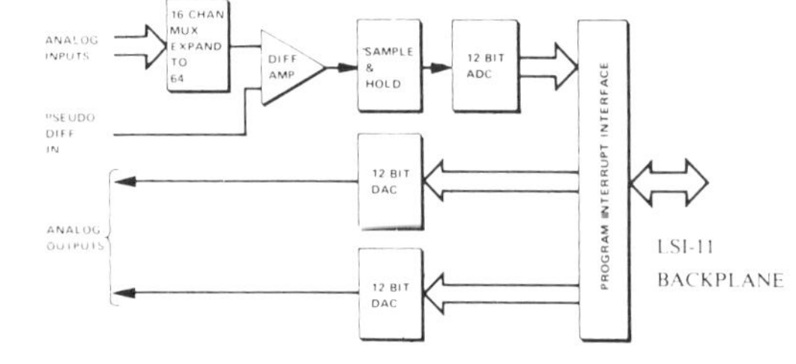

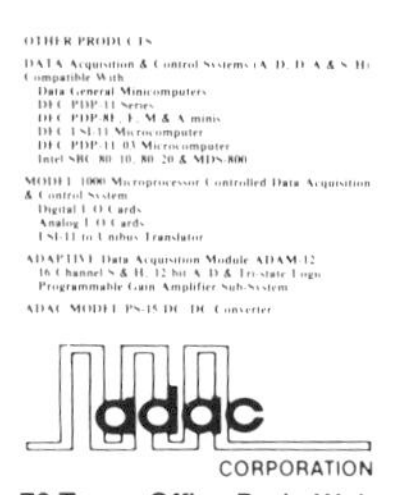

70 Tower Office Park, Woburn, MA 01801
Telephone (617) 935-6668 • Telex 949329

Fig. 7.6. Example specification sheet of an A/D and D/A card for the PDP-11/03. Reproduced with permission from the ADAC Corporation, Woburn, Massachusetts.

SPECIFICATIONS

ANALOG INPUTS
Number of Inputs to Multiplexer:
1030-16 SE: 16 Single-ended or
16 PD: Pseudo-differential
8 DI: or 8 differential
1030-32 SE: 32 Single-ended or
32 PD: Pseudo-differential
16 DI: or 16 differential
1030-64 SE: 64 Single-ended or
64 PD: Psuedo-differential
32 DI: or 32 differential
Input Voltage Range (Full Scale Range):
Standard
-10 to +10V, 0V to +10V, -5 to +5V,
0V to +5V†
With Programmable Gain
Standard ranges preceded by gains of
1, 2, 5 & 10 or 1, 2, 4 & 8
Maximum Input Voltage for Proper Operation
(Signal Plus common mode): ±10.3 Volts
Input Impedance: >100 Megohms

ACCURACY
Resolution: 12 bits
Relative Accuracy: ± 0.025% of FSR; ± .035% of FSR,
± 100 microvolts, referred to input, with
prog. gain
Inherent Quantizing Error: ±1/2 LSB

STABILITY
Tempco of Linearity: ≤3ppm FSR/°C.
Tempco of Gain: ≤30 ppm FSR/°C.
Tempco of Offset: 0.001% FSR/°C

SIGNAL DYNAMICS
Maximum Throughput Rate (12 bits): 35,000/100,000
channels/sec.
Sample & Hold Aperture Uncertainty: 20 nsec.
Crosstalk: 80db down at 1KHz "off" channels to
"on" channel
Differential Amplifier CMRR: 70dB (DC-1KHz)
Maximum Error for F.S. - F.S. Transition
Between Successively Addressed Channels: 1 LSB
Sample & Hold Feedthrough: 80db down at 1 KHz

ANALOG OUTPUTS
Number of Outputs: 0, 1 or 2
Full Scale Range:
-10V to +10V; 0V to +10V
-5V to +5V; 0V to + 5V
Impedance: <<0.1 ohms @ DC
Load Current: 5mA

ACCURACY
Resolution: 12 bits
Relative Accuracy (Linearity): ±0.012% of FSR

STABILITY
Total Output Drift at Zero Volts Output:
20ppm FSR/°C, max.
Total Output Drift at Full Scale
(Includes offset, range, linearity and
reference drift); 40ppm FSR/°C, max.

SIGNAL DYNAMICS
Settling Time to 1/2 LSB: 5 microsec.,
typical; 10 microsec., max.
Slew Rate: 10V/microsecond
Load Capacitance: 1000pF, max. for specified
settling time

ENVIRONMENTAL & PHYSICAL
Operating Temperature: 0° to 55°C
Storage Temperature: -25°C to 85°C
Size: 8 1/2" x 10" x 0.375", max.
Power:
Model 1030-X-X-X-X-X-P +5V ± 10% @ 2.5 Amps.
Model 1030-X-X-X-X-X-O +5V ± 10% @ 1.5 Amps.
+15 ± 3% @ 60 MA
−15 ± 3% @ 60 MA

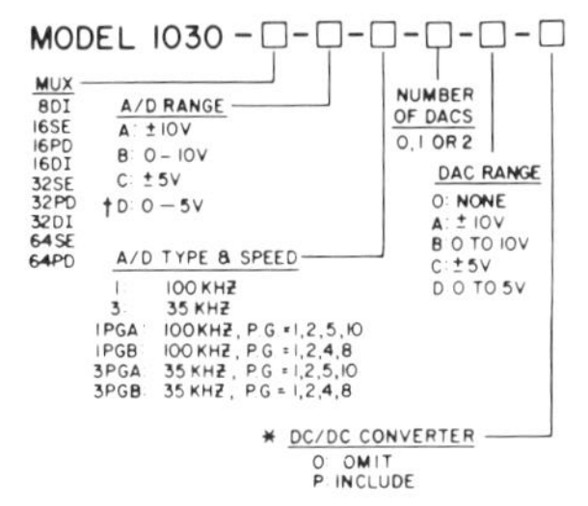

*When used with the ADAC 1000 Series a DC/DC converter is not required as ±15 volts is supplied to the backplane from system supply. When used with the DEC supplied backplanes, the DC/DC converter is required to supply the ±15 volts for the Analog circuitry.

†Not Available on 100 Khz version.

DS 1278ADA

Fig. 7.6. (continued)

The register addresses listed above may be selected to be any addresses the user desires so long as they do not conflict with other devices already allocated in the range 28K (base 10) to 32K (base 10).

The display methods using the two D/As are simpler than the A/D methods and will be presented next. Information on A/D sampling will follow as well as integrated examples.

7.4 USING THE D/A CONVERTERS

The two D/A converters available on the model 1030 can be tested by the user with virtually no preparation. They are each twelve bits thus spanning a ±2048 number range. Although their output range is selectable, we shall consider a ±5-V output for congruency with the A/D ±5-V input choice that is subsequently discussed. The addresses employed for the D/As on the model 1030 are

D/A 1: 0176760;
D/A 2: 0176762.

Figures 7.7a and b show examples of using 'as' to plot a line on the scope. In the first example, the x range is only ±400 (base 8), not very

```
        DAC1 = 0176760
        DAC2 = 0176762

plot:   mov $1000,r5
        mov $-400,r4
loop:   mov r4,DAC1
        inc r4
        dec r5
        bne loop
        br plot
```

a. draw a horizontal line program

```
plot:   mov $1000,r5
        mov $-4000,r4
loop:   mov r4,DAC1
        add $10,r4
        dec r5
        bne loop
        br plot
```

b. Expanded horizontal scale.

Fig. 7.7. Examples of 'as' programs and a 'C' program for plotting.

```
begin:  mov $-400,r2 / x axis
        mov $list,r1 / y axis
loop:   mov (r1)+, DAC1
        mov r2, DAC2
        inc r2
        cmp $list+1000,r1
        bpl loop
        br begin
list:   .=.+1000 / list with data in it
```

c.Plotting points from a list.

```
#define DA 0176760

struct {
    int     da1,
            da2;
};

main () {
    int     i,
            j;
    while (1) {
        j = 0;
        i = 0;
        while (j < 500 && i < 500) {
            DA -> da1 = i++;
            DA -> da2 = j++;
        }
    }
}
```

d.Plotting a line using C.

Fig. 7.7. (continued)

useful for being able to see data easily on the scope. Figure 7.7b shows how to expand the scale by displaying only every 10th point to allow coverage of the entire screen. (+5 V = +2047 bits.)

Timing considerations must be understood. We refresh the screen over and over to view the data. The refresh rate must be greater than about 30 times/sec for the image to appear stationary. A display of 512 points results in about 100 sweeps/sec while 4000 points will allow only approximately 12 sweeps/sec. That is,

1. 512 points × ~20 μsec/loop = ~10 msec or 100 sweeps/sec,
2. 4000 points × ~20 μsec/loop = ~80 msec or 12 sweeps/sec.

The first image appears stationary, but the second flickers. Of course, if a storage scope is used for display, speed is not an issue since no refresh is necessary. Figure 7.7c shows a similar example that allows plotting of points from a list. In 'C', plotting is accomplished by plugging values directly into the D/As. Figure 7.7d shows a simple program that draws a straight line at 45°, between (0 V, 0 V) and (1.25 V, 1.25 V). The endless loop in the while statement simply increments i and j until each reaches 500, whereupon each is reset to zero. A structure is defined for accessing the two sequential memory locations used for the D/A. Thus

```
DA -> da1;
```

refers to 0176760 and

```
DA -> da2;
```

refers to 0176762.

```
START,  TAD Y
        JMS SCALE
        CMA
        TAD X
        DCA X
        TAD X
        DILX
        JMS SCALE
        TAD Y
        DILY
        DCA Y
        JMP START

SCALE,  0
        DCA TEM
        OSR
        CIA
        DCA C
        TAD TEM
        CLL
        SPA
        CML
        RAR
        ISZ C
        JMP .-5
        JMP I SCALE
X,      1111
Y,      2222
C,      0
```

Fig. 7.8. Kaleidoscope program for the PDP-8. This program uses the switch register to obtain a "seed" for a random number. As the least-significant bits on the switch register are changed, the display will assume a variety of complex patterns, sometimes static, sometimes dynamic.

7.5 KALEIDOSCOPE

The kaleidoscope program is a fascinating small display program. Originally written for the PDP-8 and provided for the Lab 8/e by Digital Equipment Corporation, the PDP-8 version shown in Fig. 7.8 was recoded in PDP-11 code (Fig. 7.9). In the original PDP-8 version, values are put

```
.=.+1000

begin:  mov pc,sp
        tst -(sp)                   /set up the stack
        bis $01, *$177560           /enable keyboard
        bic $100,*$177564           /turn off interrupt
start:  mov y,r2
        jsr r2,scale
        com r3
        add r3,x
        mov x,r3
        mov x,*$176760              /D/A one
        jsr r2, scale
        add r3, y
        mov y,*$176762              /D/A two
        br start

scale:  mov r4,c
        tstb *$177560
        bmi get
continue:mov (sp),r3
loop:   clc
        tst r3
        bpl next
        sec
next:   ror r3
        inc c
        tst c
        beq return
        br loop
return: rts r2

get:    mov *$177562,r4
        bic $0177740,r4
        neg r4
        mov r4,c
        br continue

x:      1111
y:      2222
c:      0
```

Fig. 7.9. An 'as' version of the PDP-8 program in Fig. 7.8.

into the x and y D/A outputs by use of the instructions DILX and DILY (display X and display Y). The values used for the ''seed—the random values used to select a new pattern'' were entered from the switch register by the command OSR ('or' with switch register). However, in the 11/03 there is no switch register; thus, the method for entering an initial 'seed' was changed to provide input from the terminal (see 'get').

The interested reader should compare the PDP-8 and PDP-11 versions of this display routine. The PDP-8 instructions are

TAD Y	:	two's complement add. The value Y is added into the accumulator (like a register)
CMA	:	complement
DCA Y	:	deposit contents of accumulator into Y
CIA	:	negate
CLL	:	clear link (the carry bit on the 8)
JMS	:	jump to a subroutine
DILX	:	Display X D/A
DILY	:	Display Y D/A
SPA	:	skip on positive accumulator

What is interesting in this translation is the relative ease in converting PDP-8 code to PDP-11 code. Can you see the similarities between the two sets of code? Would you be able to translate other PDP-8 code to PDP-11 code?

7.6 SAMPLING SPEED

The Nyquist sampling theorem (Bendat and Piersol, 1971) states that one must sample at a frequency at least twice the highest frequency present in the data. If this condition is not met, *aliasing* will occur. That is, the sampled signal can appear the same as another signal of different fre-

Fig. 7.10. A sine wave with sufficient sampling points to avoid aliasing.

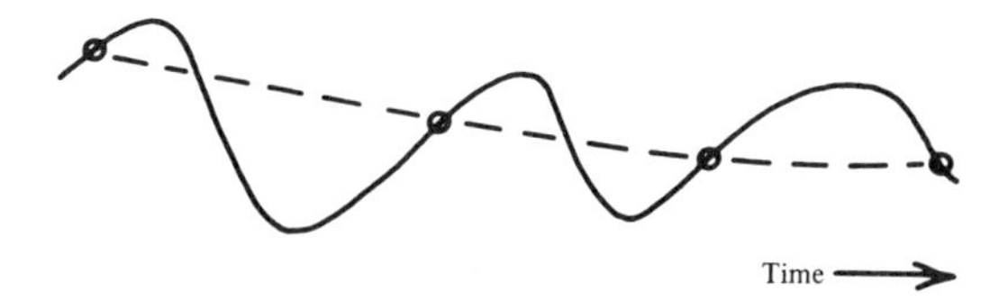

Fig. 7.11. A sine wave with an insufficient sampling frequency.

quency or have an *alias*. For example, consider the signal shown in Fig. 7.10. Sampling the signal wherever there is a 0 will produce a clear digital representation of the original signal. But, if we sample insufficiently rapidly, e.g., as shown in Fig. 7.11, the digital signal could have been obtained by sampling an analog signal with the shape of the dotted line, which is not at all like the original signal. Thus, it is incumbent on the user of the A/D system to limit the bandwidth of the input signal and to specify accurately the sampling rate. A common practice is to use analog filters before the A/D inputs to ensure that no signals can appear at the A/D input higher than twice the sampling rate. A typical choice for a sampling rate is three to four times the highest frequency that is expected to be found in the data.

An easily understood demonstration of aliasing occurs when wagon wheels in western movies appear to move backwards, stand still, or move at various forward rates instead of the way they are observed to move with the naked eye. These observable not-true-to-nature effects are due to the aliasing phenomenon of inadequate sampling. The wagon wheel is simply moving too fast to be realistically represented by presentation of 30 frames/sec of the scene.

7.7 A/D SAMPLING

The A/D uses two registers, the status and data registers. Table 7.1 shows the bit allocations in the status register. Setting bits in this location (0176770), will allow the user to

1. start the A/D,
2. enable an external source to start the A/D,
3. set the sequential or random mode for stepping through the channels,
4. set the gain,
5. set the channel, and
6. enable the interrupt.

TABLE 7.1

BIT SETTINGS AND DESCRIPTION OF A/D CONTROL/STATUS REGISTER OF THE MODEL 1030 DATA ACQUISITION AND CONTROL SYSTEM[a]

Bit	Signal	Address 176770 Description
D15	Error	Set if ADC trigger occurs and previous conversion is not complete. Interrupt is produced when interrupt bit (D6) is enabled.
D14	Self-test	Used for maintenance purposes only.
D13	MUX-channel 2^5	
D12	MUX-channel 2^4	
D11	MUX-channel 2^3	Loads multiplexer address to select one of 64 channels and initiates a conversion (if EXT Enable, D1, is a zero)
D10	MUX-channel 2^2	
D9	MUX-channel 2^1	
D8	MUX-channel 2^0	
D7	Done	Set by completion of conversion and reset upon reading data register or initialize.
D6	Int. Enable	Program selectable interrupt mode. Interrupt produced by ADC done (D7) or error (D15) when selected.
D5	Reserved	Used for special applications only.
D4	Gain 2^1	Sets gain of programmable gain amplifier option. 11 sets lowest gain and 00 sets highest gain.
D3	Gain 2^0	
D2	Seq/Rand	Zero selects random mode for multiplexer. One selects sequential mode for multiplexer. In sequential mode, multiplexer register is automatically incremented at end of each conversion. Triggering of ADC is same as in random mode.
D1	Ext. Enable	Enables clock source to trigger ADC. Jumpers select on-board multivibrator or external trigger.
D0	Start	Triggers ADC, if Ext. Enable, D1 is a zero.

[a] Reprinted with permission of ADAC Corporation, Woburn, Massachusetts.

Also, one bit (#7) is reserved to tell the user when sampling is complete and another bit (#15) to indicate if an error occurred during sampling. After a sample is taken, the digital value appears in the data register (0176772). The value in this location can then be moved to another location. The actual numerical values of the sampled value can be estimated as shown in Fig. 7.12 for an A/D range of ±5 V for a 12-bit A/D. Consider the point on the sine wave marked with an X in Fig. 7.12. At an input level of 2.5 V, the digitized value of the sample at this amplitude would be about

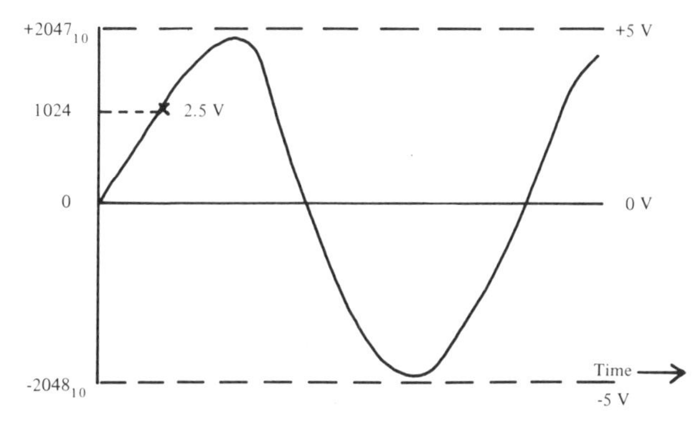

Fig. 7.12. Voltage levels relative to binary digitized values for a 12-bit ±5-V A/D.

1024 (base 10) or 2000 (base 8). If this sample was examined with ODT, it would appear in memory as 02000. Other values on the sine wave shown can be similarly computed. The reader should be aware of the allowable input signal levels and the voltages physically represented by the values sampled.

Example and exercise. On the 11/03 type

```
@ 176770/. . . . . . . 01
@ 176772/xxxxxxx
```

Determine what xxxxxxx is. What have you done? Now, attach a voltage of 1 V DC to A/D channel 0 and repeat the experiment. (Caution, be sure it is only 1 V, not greater than +5!) What is the result?

7.8 METHODS FOR SAMPLING

Three basic methods for sampling are available with a number of possible variations in the basic categories. Figure 7.13 conceptualizes in flow chart form the three methods:

1. *Wait loop method.* In the wait loop technique, after the A/D is started, we wait for 20 μsec until sampling is completed and then read the value that appears in the data register. Until it is time for the next sample, iterations of a wait loop are made.

2. *Clock driven sampling.* As shown in Fig. 7.13b, the A/D is set to wait until the clock sends a signal that a sample should be taken. Various

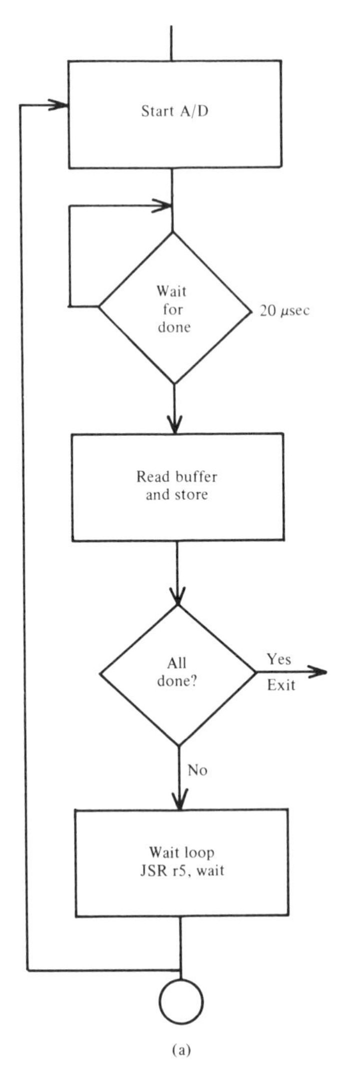

Fig. 7.13. Methods for sampling. (a) Wait loop, (b) clock wait, and (c) Interrupt method.

possibilities exist for use of a clock or other external drivers. For example,

(a) A real-time clock can be used. Clocks are often provided on separate cards that can be programmed to tick (count) at rates up to usually 1 MHz.

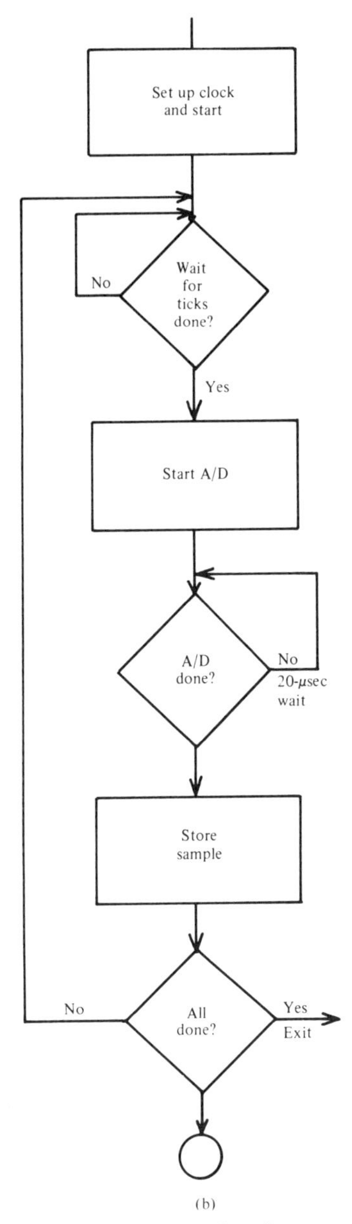

(b)

Fig. 7.13. (continued)

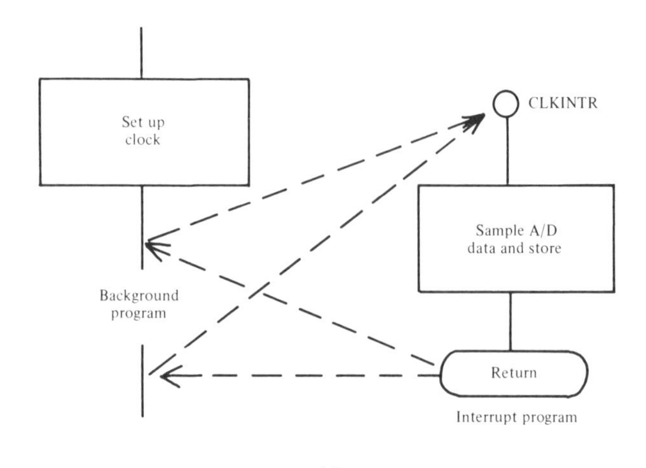

Fig. 7.13. (continued)

(b) A simple multivibrator might be employed. For example, a 555 IC timer chip. A 555 is included on the model 1030 board.
(c) A switch that you can press.
(d) The 60-Hz clock in the computer.
(e) A signal generator that will also serve to trigger the A/D.

The external trigger input is often preceded by Schmitt trigger circuitry so any shape waveform can be used for triggering. Some clock cards are available with several Schmitt trigger inputs that have variable slope detection and level setting. Such versatility is frequently useful when the input signals employed are not uniform in nature.

3. *Interrupt driven.* Figure 7.13c depicts the method employed when the computer's interrupt system is turned on. The clock is set to interrupt at a specific rate. At each interrupt, the interrupt service routine (ISR), is executed and a sample is taken. After completing the isr, the program returns to the program it was previously executing (see Chapter 8 for a complete discussion of interrupts).

7.9 EXAMPLES

Consider the 'as' program shown in Fig. 7.14 that does nothing more than start the A/D, sample, and wait.

```
=.+1000                             /header of 1000 for RT-11

mt = 104000                         / used for RT-11 exit trap
dstat= 176770
ddat=  176772

dtest: mov      $500., r0           /500 samples
       mov      $list,r1            /set up address to start storing
                                    /sampled data
ept:   mov      $01,adstat
       mov      $3,r4
dloop: tstb     *$adstat            /finished sampling A/D ?
       bpl      wdloop              /no,then wait until finished. doneflag=1
       mov      *$addat,(r1)+       /store data in buffer
       dec      r0                  /finished taking 500 samples?
       bne      waitlp              /no, then wait~~ approx. 500msec
                                    /between samples
       emt ¦ 350                    /yes,then exit using trap to RT-11 or halt
                                    /use 'sys exit' in MiniUNIX
aitlp: mov      $21643.,r3          /execution time on 11/03 = 4.9 microsec
p0:    dec      r3                  /4.2 microsec
       bne      lp0                 /3.5 microsec
                                    /the total instruction time for the
                                    /loop is 7.7microsec-> 7.7*21643=
                                    /166651.1microsec
       dec      r4                  /this loop takes 7.7*3=23.1 microsec
       bne      waitlp
       br       rept                /if time loop finished branch
                                    /to get next sample

total wait time= (21643*7.7*3)+(4.9*3) + 23.1 +3.5 =499994.6 microsec
this loop is 5.4 microsec under the 500 msec desired.

ist:   .=.+500.
```

Fig. 7.14. A/D sampling using an 'as' program. This program samples data from the A/D with approximately 500 μsec between samples. The delay between sampling points is accomplished by a wait loop. Thus, the time delay is not exactly 500 μsec.

1. 500 is moved to r0 and subsequently used as a counter for the number of samples to be taken. "dec r0" occurs in each loop until r0 reaches 0 and the program exits.
2. "mov $list,r1" moves the address of list into r1. Then, when r1 is used, it is employed as a pointer to indicate where the data are stored. list: .=.+500. reserves 500 decimal locations in which data can be stored.
3. "mov $01, adstat" starts the A/D by putting a 1 in the status register. Note that an equivalent statement is

```
mov $01, *$0176770
```

In using the first method, we employ PC relative mode. An offset from the current location to the address of adstat will be stored following the mov instruction. In the second case we indirectly access the location via immediate storage of the value in the second word location following the "mov."

4. "tstb *$adstat" tests the byte in location 0176770. If a 1 appears in bit 7 (i.e., the byte is negative), then the test produces a flag (=1) in the PSW! Thus, we bpl to wdloop until the flag is raised. A typical time in this loop is 20 μsec.

5. Data are stored by moving the contents of 0176772 (addat) to list in the instruction "mov *$addat, (r1)+."

6. Next, we either exit or wait. After waiting we branch back to "rept:" and take another sample. Looping continues for 500 iterations and the list of values sampled is stored in list.

Consider the wait routine that begins at "waitlp:." Waiting can consist of simply a series of do-nothing instructions that just burn up time. In this example, r3 and r4 are used as counters and decremented with the dec instruction until they equal 0. The program comments document the computation of the length of the wait (approximately 0.5 sec in this example).

7.10 A/D SAMPLING IN 'C' USING A WAIT ROUTINE

Figure 7.15 displays a program in 'C' that is conceptually equivalent to the 'as' program in Fig. 7.14. The program is somewhat longer, but is easier to read. Structures are defined for the A/D and D/A and definitions for the device registers given (a later example will demonstrate how this technique lends clarity to the program). An added feature of this program is that it will display the sampled data, both during sampling and after completion of sampling until halted (press 'break' or include code to exit gracefully). The variable "p" points to integers and, "int buff[500]" defines 500 locations for values (note here they are automatic, i.e., go on stack). Initially, "p = buff" sets p equal to the address of buff[0]. We start the A/D, wait for it to finish and put the sampled value in buff[ ]. Take careful note of the while statement and the equation

```
*p = AD->buffer;
```

This statement simply means put in the address in p the value in buffer (in this case the value in location 0176772). The value stored in buff[ ] is displayed in da1 and p is then incremented so that the address points to the next location. Since only one D/A is used, the scope internal sweep can be

```
*

#define AD 0176770
#define DA 0176760
#define ADDONE 0200
#define START 01                        /* start A/D on channel 0 */

struct {
    int     status,
            buffer;
};
struct {
    int     da1,
            da2;
};
main () {
    int     i,
            k,
           *p,
            j,
            buff[500];                  /* reserve 500 words for buffer */

    p = buff;
    for (i = 0; i < 500; i++) {
        AD -> status = START;
        while (!(AD -> status & ADDONE));
                                        /* wait for done bit */
        *p = AD -> buffer;              /* store sample away in buffer */
        DA -> da1 = *p++;               /* display the point and increment
                                           */

        wait ();                        /* wait between samples */
    }
    while (1) {
        p = buff;                       /* display when done sampling */
        for (i = 0; i < 500; i++) {
            DA -> da2 = i;              /* drive sweep with increment of i
                                           */

            DA -> da1 = *p++;           /* display data on y axis of scope
                                            */

        }
    }
}
wait () {
    int     i;
    for (i = 0; i < 500; i++);          /* wait five hundred iterations */
                                        /* doing nothing */
}
```

Fig. 7.15. An A/D sampling routine written in 'C' that uses a wait loop between samples.

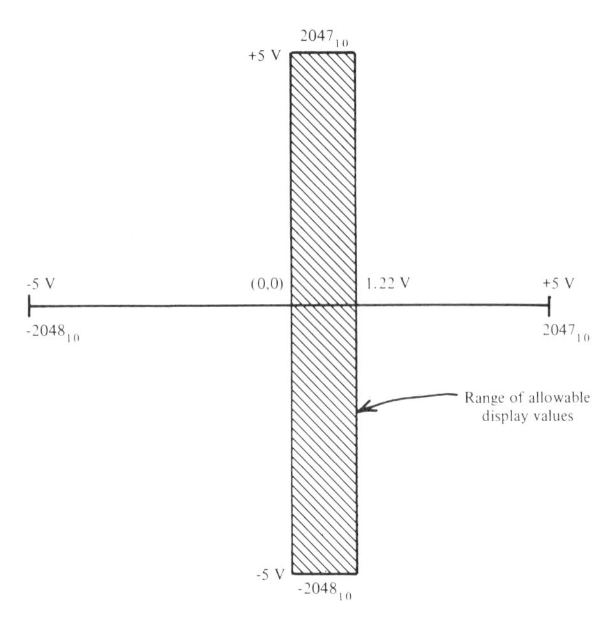

Fig. 7.16. Display constraints in Fig. 7.15. The cross-hatched area represents allowable display voltages produced by the program in Fig. 7.15.

used to display data points across the screen as they are being sampled. After sampling, the entire sampled data buffer (buff[ ]) is displayed over and over again using da2 to drive the x axis on the scope. "wait( )" is a function call. Note that wait( ) is outside main and shares no variables (i is automatic).

The display output in Fig. 7.15 is not optimal. Consider the voltage range over which the x and y inputs can range. The y axis produces the same range of values as were originally sampled; i.e., usually ± 5 V for full scale. However, the x-axis values provide a display only from 0 to 1.22 V as shown in Fig. 7.16, a dramatic underutilization of the horizontal width of the scope face. Naturally, the x amplifier of the scope can be adjusted to make the sweep appear across the full scope width. However, it would be good practice to standardize on a common method for x-axis generation so that the scope does not have to be adjusted for every different program.

7.11 STYLE CONSIDERATIONS

Figures 7.17a and b show two methods for writing a program for displaying a line on the oscilloscope. Figure 7.17a is written in the style of

```
#define DA 0176760

int     i;

struct {
    int     da1,
            da2;
}

main () {
    for     (i = 0; i < 500; i++) {
        DA -> da1 = i;
        DA -> da2 = i;
    }                                   (a)
}
```

```
main () {

    int     *p,
            i;
    p = 0176760;
    for (i = 0; i < 500; i++) {
        *p = i;
        *(p + 2) = i;
    }
}                                   (b)
```

Fig. 7.17. Two programs to draw a line: (a) conventional style with a structure, and (b) direct register access using a pointer.

```
main () {
    int     i,
           *p,
            buff[500],
           *bp;
    bp = buff;                          /* set bp to address of buff */
    p = 0176760;                        /* set p to D/A LOC */
    for (i = 0; i < 500; i++) {
        *(p + 10) = 01;                 /* AD status = 0176770 = p+10 */
        while (!(*(p + 10) & 0200));    /* 01 = START */
        *bp = *(p + 12);                /* AD to buff !! */
        *p = *bp++;                     /* p points to D/A - therefore,
                                           buff D/A */

        wait ();                        /*        is displayed */
    }
    while (1) {
        bp = buff;
        for (i = 0; i < 500; i++) {
            *(p + 2) = i;
            *p = *bp++;
        }
    }
```

Fig. 7.18. Revision of Fig. 7.15 using pointers rather than structures. Note confusing appearance.

Fig. 7.15 while Fig. 7.17b is written more tersely. Convince yourself that the two programs should actually do the same thing. Next, read Fig. 7.18 and compare it with Fig. 7.15. The code in Fig. 7.18 is written in the style shown in Fig. 7.17b and should be equivalent to the code in Fig. 7.15. The ability of a reader to figure out what this code does in Fig. 7.18 is seriously hampered by the style of writing. This example is given to convince you to follow a standard and clear style rather than use the most obscure coding arrangements you can find.

7.12 USE OF THE REAL-TIME CLOCK

Each PDP-11 CPU contains a line-time clock (60 Hz), which can be accessed by the user on the 11/03. However, it is an infrequent occurrence that the user wishes to sample at 60 Hz or use the clock for other activities at exactly 60 Hz. Thus the user normally purchases a *real-time clock* that will allow flexible timing. This section describes how to use a specific real-time clock, the KW11-P, and gives two examples.

The KW11-P is essentially the same as the KWV11-P for the PDP-11/03. Cards made by other manufacturers, such as MDB systems, also have similar programming characteristics.

Table 7.2 shows the status register configuration for the KW11-P reproduced from the DEC Peripherals Manual. The control and status register is similar to the typical control and status configuration of other devices. Rate is selectable in bits 1 and 2. Experimentally, it was observed that the bit settings for the KW11-P shown in Table 7.2 did not function correctly on several clock cards tested. When the bit table was modified to be as shown in Table 7.3, the rates operated as specified. The above rate table may well be different on different hardware versions of the KW11-P. Users should carefully check the characteristics of hardware they employ prior to including code for the hardware in complex and difficult-to-debug programs.

The clock has three registers as shown in Fig. 7.19. The location labeled

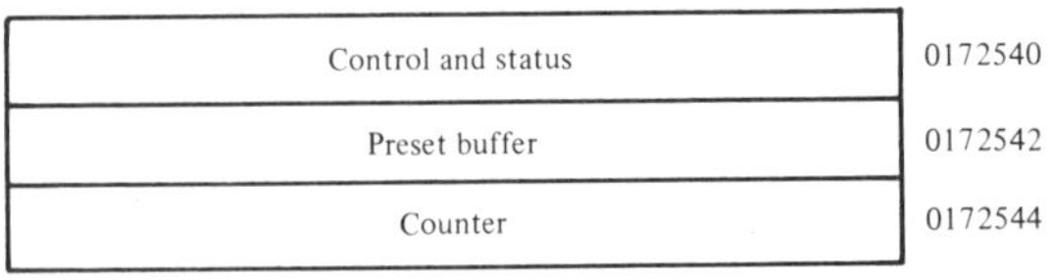

Register	Address
Control and status	0172540
Preset buffer	0172542
Counter	0172544

Fig. 7.19. Three-word device status register configuration for the KW11-P real-time clock.

"preset buffer" is also often called the counter-set buffer register. The *fundamental concept* is that one can obtain virtually any time length between clock interrupts by counting either up or down at a specified rate (e.g., 10 kHz or 100 kHz) for a specified number of counts until either overflow or underflow occurs. Thus with a 16-bit counter register a maximum of 2^{16} or 64K clock ticks can be counted before under or overflow occurs. Bit 7 (the done bit) is set when either underflow or overflow occurs. The counter register is a binary up/down counter. It is read only and should be read before stopping it. Bit 3 (mode) in the cs register sets either single or repeat interrupt mode. What this means is that the clock will interrupt (set-done flag, if the interrupt flag bit is not enabled) only once when bit 3 = 0, but when bit 3 = 1 it will transfer the contents of the preset buffer into the counter and begin counting again after each interrupt.

Suppose we wish to produce a signal (i.e., a flag to be tested or cause an interrupt to occur) every $\frac{1}{64}$th of a second. First, let us select the clock rate to be 100,000 Hz. Selecting bits as shown below

bit 0 = 1	Start
bit 1 = 0	100 kHz
bit 2 = 0	
bit 3 = 1	Repeat
bit 4 = 0	down

selects start, downcount, and repeat mode at 100,000 Hz. Thus 1562.5 counts are required to produce a $\frac{1}{64}$-sec delay, i.e.,

$$100{,}000/64 = 1562.5$$

We cannot, however, have half counts so we must choose either 1562 counts or 1563 counts. Choosing either one gives a sampling rate slightly different from 64 Hz, i.e.,

$$100{,}000/1562 = 64.02048656$$

or

$$100{,}000/1563 = 63.9795265$$

In 100 sec of sampling, with these two rates, the actual elapsed time would be

99.968 and 100.032 sec

Before setting the clock CSR, the value computed for the number of counts must be placed in the clock preset buffer.

TABLE 7.2

REPRODUCTION OF KW11-P REAL-TIME CLOCK INFORMATION FROM THE DIGITAL EQUIPMENT CORPORATION PDP-11 PERIPHERALS HANDBOOK[a]

FEATURES

- four clock rates, program-selectable
- crystal-controlled clock for accuracy
- two external inputs
- three modes of operation
- interrupis at 50 or 60 Hz line frequency

DESCRIPTION

The KW11-P Clock provides programmed rea' ιime interval interrupts and interval counting in three modes of ' peration. The major functional units of the Clock include:

16-bit Counter: Counts up or down at four selectable rates and can be read while operating. The interrupt sequence is initiated at zero (underflow) during a countdown from a preset interval count. The count-up mode is used to count external events; an interrupt is initiated at 177 777 (overflow).

16-Bit Count Set Buffer: Stores the preset interval count. At underflow, depending on the operating mode, the buffer automatically reloads the Counter or is cleared.

Control and Status Register: Provides various control and status signals related to the operation of the buffer and counter.

Clock: Provides 2 crystal-controlled signals of the 100 kHz and 10 kHz to clock the counter. Two external clock inputs are provided: 50/60 Hz line frequency and a TTL-compatible signal input.

MODES OF OPERATION

Single Interrupt Mode: A program-specified time interval is present and an interrupt is generated at the end of the interval. The time interval, represented as a specific count, is loaded into the counter. Countdown or count up is initiated at one of four selectable rates; at underflow or overflow an interrupt is generated, clocking is stopped, and the counter is reset to zero.

the counter in the count-up or countdown mode. The counter may be read during operation to determine the number of events that have occurred.

REGISTERS

Control and Status Register 772 540

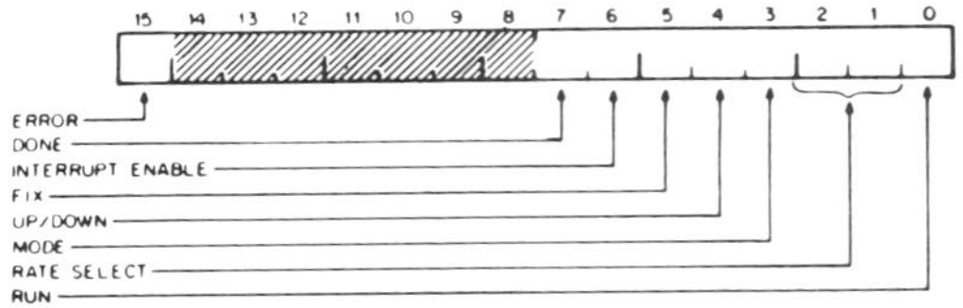

15 14 13 12 11 10 9 8 7 6 5 4 3 2 1 0

ERROR
DONE
INTERRUPT ENABLE
FIX
UP/DOWN
MODE
RATE SELECT
RUN

Bit: 15 **Name:** Error

Function: Set when, in repeat interrupt mode, a second underflow or overflow occurs before the interrupt of the preceding one has been serviced. It is cleared when the Status Register is addressed and by internal gating. It is valid only during the first serviced interrupt after the error.

Bit: 7 **Name:** Done

Function: Set on underflow or overflow.

Bit: 6 **Name:** Interrupt Enable

Function: Set to allow Done = 1 to cause an interrupt.

Bit: 5 **Name:** Fix

Function: Set to cause single clocking of the counter as a maintenance aid.

Repeat-Interrupt Mode: A program-specified time interval is present and repeated interrupts are generated at a rate corresponding to the time interval. Operation is similar to the Single-Interrupt Mode, except that after the interrupt is generated on underflow or overflow, the counter is automatically reloaded from the control-set buffer and clocking is restarted. At the next underflow or overflow, another interrupt is generated.

External Event Counter Mode: The external input is used to clock

Rate	Bit 2	Bit 1
100 kHz	0	0
10 kHz	1	0
Line frequency	0	1
External	1	1

Bit: 0 **Name:** Run

Function: Set to allow the counter to count. Cleared on underflow in single-interrupt mode.

Count Set Buffer Register 772 542

This 16-bit register is used for storage of the interval count. It allows automatic reloading of the Counter in repeat-interrupt mode. The register is cleared by the INITIALIZE signal and by underflow or overflow in the single interrupt mode. The bits are write-only.

Counter Register 772 544

This 16-bit register is a binary up/down counter. It is cleared by the INIT (initialize) signal. The bits are read-only.

PROGRAMMING

Read the counter prior to stopping it. Stopping the counter might change its contents. If it is necessary to start the counter from a previous value, save the value which was read and reload if required. Do not loop on a counter-read command.

The latest version is equipped with a hardware synchronization feature which will add from zero up to one clock interval (of the selected rate), to the anticipated count time on the first interrupt after the run bit is asserted.

Bit: 4 **Name:** Up/Down

Function: Selects either count-up or countdown for the counter; 1 = up, 0 = down.

Bit: 3 **Name:** Mode

Function: Selects interrupt mode of operation; 1 = Repeat Interrupt, 0 = Single Interrupt.

Bit: 2-1 **Name:** Rate Select

Function: Selects one of four available clock rates.

SPECIFICATIONS

Main Specifications

Clock rates	100 kHz, 10 kHz crystal-controlled, line frequency, external (Schmitt Trigger input), oscillator stability: ±0.01%
Operating Modes	Single interrupt, repeated interrupt, external event counter, non-interrupt
Register Addresses	
Control and Status	772 540
Count Set Buffer	772 542
Counter	772 544
UNIBUS Interface	
Interrupt vector address	104
Priority level	BR6
Bus Loading	1 bus load
Mechanical	
Mounting	1 SPC slot (quad module)
Power	
0.5 A at +5V	

TABLE 7.3

RATE SELECTION BY BIT SETTING ON THE KW11-P CLOCK

Rate	Bit 2	Bit 1
100 KHz	0	0
10 KHz	0	1
Line frequency	1	0
External	1	1

7.13 CLOCK EXAMPLE

Figure 7.20 displays an example of a 'C' program used to ring the terminal bell 10 times, waiting 2 sec between each bell.

START is defined as 013. Referring to the CSR description, it can be observed that this value corresponds to down counting at 10 kHz (see Fig. 7.21). Bit 0 = 1 starts the clock. Bit #3 is set to indicate that the clock is in repeat mode; i.e., each time the clock counter counts down to zero, it is automatically reset to TICKS.

"TICKS" is defined as 20000 (base 10) (note no 0 means this is a decimal number). Putting this number into the preset register

```
CLOCK->preset = TICKS;
```

will produce a down count time of 2 sec. That is, we count 20,000 items at a rate of 10,000/sec. The buffer preset register is reloaded with TICKS each time the clock counts down to zero. A while loop waits until the clock has counted down to zero, the terminal bell is rung, and the clock automatically reset. At the outset, we turn off the printer interrupt. Otherwise, the program will interrupt at the completion of the ringing of the bell. Bit 6 (0100) is set to zero to turn off the interrupt (see Chapter 8).

If the two statements

```
CLOCK->buffer = TICKS;
CLOCK->status = START;
```

are relocated inside the "for" loop and START is defined as 003 (i.e., no automatic reloading of the clock counter), the clock can be started on each pass through the for loop.

```
define CLOCK 0172540
define DONE 0200
define START 013
define TICKS 20000
define KLPRINT 0177564

* The above settings produce a down count at 10 KHz and start the clock.
 With START = 015, TICKS =120 the same result will be produced.
 The clock may be restarted each time if the automatic reset feature
 is not used.
/

truct {
   int     status,
           buffer,
           counter;
;

ain () {
   int     i;
   KLPRINT -> status =& ~0100;          /* turn off KL interrupt */
   CLOCK -> buffer = TICKS;
   CLOCK -> status = START;
   for (i = 0; i < 10; i++) {
       while (!(CLOCK -> status & DONE));
       KLPRINT -> buffer = 07; /* bell = 7 */
   }
```

Fig. 7.20. Example of using the KW11-P real-time clock to ring periodically the bell on the console terminal.

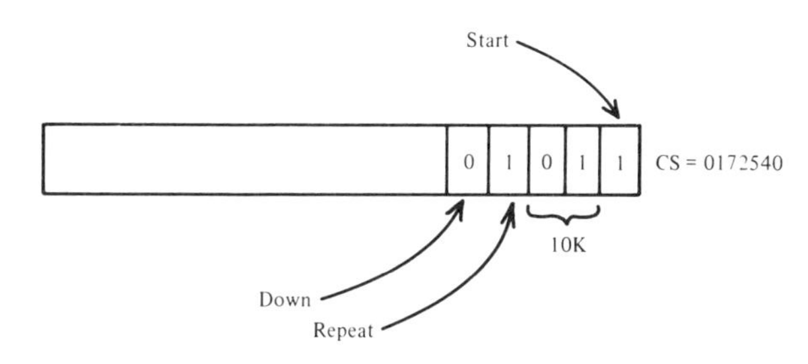

Fig. 7.21. Specific setting of the control status register in Fig. 7.20.

7.14 CLOCK AND A/D CONVERTER

It is a relatively short step from the previous example to constructing an example for driving the A/D with the clock. Consider replacing the bell with starting the A/D. Figure 7.22 shows the basic program. Definitions and structures are the same as in the previous example. Here, they are included by the

#include "defs.h"

statement. By setting START = 011, repeat mode and 100 kHz for the clock are used. In this example, 500 samples are to be taken and stored in buf[ ], and then displayed on the oscilloscope. A pointer p is defined to point to integers (int *p;) and set equal to buf (the first location in buf)

```
#

#include "defs.h"

int     buf[500];

main () {
    int     i,
           *p;
    p = buf;
    CLOCK -> buffer = 1562;      /* number of ticks */
    CLOCK -> status = START;
    for (i = 0;i< 500; i++) {
        while (!(CLOCK -> status & CLOCKDONE));
        AD -> status = START;
        while (!(AD -> status & ADDONE));
        *p++ = AD -> buffer;
    }
    CLOCK -> status = 0;         /* turn off the clock */
    display ();
}

display () {
    int     i,
           *p;
    while (1) {
        p = buf;
        for (i = -2000; i <= 2000; i =+ 8) {
            DA -> da1 = i;
            DA -> da2 = *p++;
        }
    }
}
```

Fig. 7.22. Example of using the clock to control the sampling time interval.

before sampling begins. After each sample is taken, it is stored away in buf (*p++) and incremented to point to the next location. "display( )" is called to display repetitively the contents of buf. In this example, pointers are used to access the buffer. One can also use array addressing.

EXERCISES

1. Halt the PDP-11/03 and type

   @0176760/. . . . . 1000 CR
   @0176762/. . . . . 1000 CR

   What happens on the oscilloscope? Why? Can you verify that the beam on the scope moves as you expect when you enter different values into these two registers? (It should be noted that the outputs of the two D/A must be plugged into the x and y inputs of the scope for this experiment to work successfully!)
2. Compute the exact coordinates for the line plotted in Fig. 7.7d.
3. Examine the .s file of the 'C' program in Fig. 7.7d. How do the values for i and j get placed in 0176760 and 0176762?
4. What is the speed of a D/A? Are there any constraints? If you do not know, can you devise a method for finding out?
5. Store a sine wave in a linear array and display on the scope so that the display appears stationary.
6. Run the 11-based kaleidoscope shown in Section 7.5 and observe the patterns produced.
7. Suppose the output of an amplifier produces a calibration signal of the form shown in Fig. 7.23 when a $\pm 50\ \mu V$ square wave signal is fed into the input (Fig. 7.24).

   (a) Determine the gain of the amplifier.
   (b) For this calibration, what would be maximum and minimum sampled values in octal and decimal for a ± 10-μV input sine wave? Produce a hypothetical or actual list of numbers representing the A/D sampled data assuming: ± 5-V A/D input window, 10-Hz input signal and a 50-Hz sampling date.

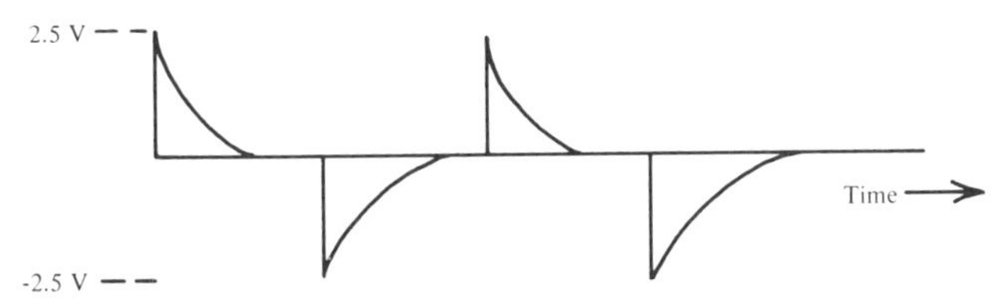

Fig. 7.23. Calibration signal output.

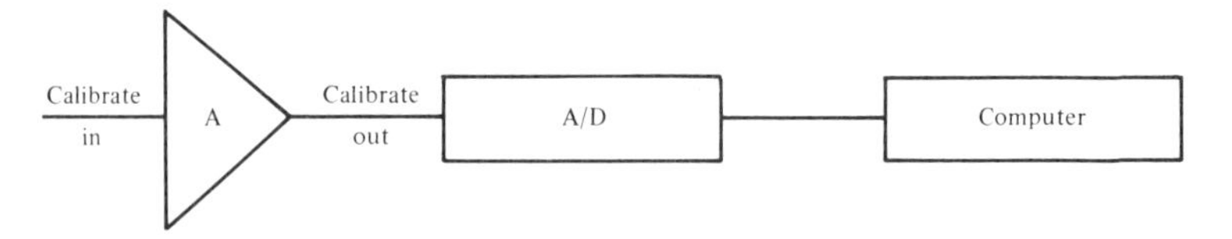

Fig. 7.24. Amplifier connected to A/D.

8. List all conditions that are set when $01 is put into the status register of the A/D. Write a wait loop in 'as' to produce a 1-sec delay.
9. Code the program in Fig. 7.14 by hand and run with ODT.
10. Code the program in Fig. 7.14 in Macro and 'as' and run.
11. Examine the contents of list when you run the program in Fig. 7.14 with the signal you put in (yes, you have to put one in). Explain the numbers you have obtained in list versus what you fed into the A/D.
12. Explain how to store the sampled values more efficiently than in the problem above. You are now wasting the upper 4 bits of each word. Consider (!) (1) using these 4 bits for storage in some way, (2) byte packing, and (3) differential encoding. Differential encoding refers simply to storing the difference between each sample and byte packing means storing two bytes per word. Describe the results you would expect using different methods for improving the efficiency of data storage.
13. In Fig. 7.15: (1) Calculate how long wait( ) waits! Use the .s file to make your determination. (2) In the while loop for display, when p is set equal to buff on each iteration, what is being done?
14. Modify the display in Fig. 7.15 so that the x axis will sweep from -5 to $+5$ V in 500 steps. Next, modify this program to use the clock rather than the wait( ) function. Does it make any difference what clock rate is used?
15. See what difference using pointers versus arrays makes in the program in Fig. 7.22. Write one version with no pointers and print out the corresponding .s file. Repeat with pointers. Analyze the two sets of code. What are the differences in speed for the two versions?
16. Determine the maximum sampling speed possible given a 20-μsec settling time for the A/D. Use your .s file to compute this time. Are there different ways to improve speed?
17. Suppose you wish to sample two channels at the fastest rate possible. What is this rate and what is the minimum time between channel sampling?

REFERENCES

Bendat, J. S., and Piersol, A. B. (1971). "Random Data: Analysis and Measurement Procedures." Wiley (Interscience), New York.

Rabiner, L. R., and Gold, B. (1975). "Theory and Application of Digital Signal Processing." Prentice Hall, New Jersey.

8

Interrupts and Real-Time Programming

8.1 INTRODUCTION

In a real-time environment, peripheral devices that perform time-critical tasks must be serviced without delay. Thus any running program must be *interrupted* to allow a device to collect data or perform other tasks. After executing a series of instructions associated with the real-time event, the program is permitted to resume execution wherever it was interrupted. For example, if an A/D channel is to be sampled 100 times/second, each $\frac{1}{100}$ of a second the A/D must be serviced (a sample taken and stored away). Hence, whatever program is being executed is *interrupted* momentarily to service the A/D. Almost every DEC peripheral device is provided with the ability to interrupt by setting (usually) bit 6 of the control/status register of the device.

There are significant advantages to using the interrupt system. In the above example, we could wait by using a loop in the program between each sample for $\frac{1}{100}$ of a second. Yet, waiting in a loop wastes time that instead could be used to perform useful tasks or computations. To provide an estimate of the amount of time available, assume that taking one sample requires on the order of 100 μsec. Thus, if 10 msec is the time between samples, there are about 9.9 msec available that are unused. In this time a significant number of instructions can be executed. Assuming an average of 5 μsec/instruction, there are around 200 instructions/msec $\times$ 9.9 = 1980 instructions that can be executed between samples. This number of instructions is clearly enough to perform a large number of computations in the given time period. In fact, during a 100-sec sampling period in which samples are taken at a rate of 1 sample/sec, only 10 msec (100 $\times$ 100 μsec) are used to perform the actual sampling. Based on this observation, it is obviously critical to design programs efficiently so that interrupts can be used in most real-time situations.

8.2 INTERRUPT MECHANISMS

Different computers often use different methods for performing interrupts. We shall describe the methods used for the PDP-8 and PDP-11. The first machine uses a technique known as a "skip chain" and the second machine employs a "vectored" interrupt.

8.2.1 PDP-8 Interrupts

The following description is primarily conceptual and is intended as a set of remarks to contrast with the PDP-11 discussion that follows.

In the PDP-8, bits are set in device registers to allow devices to interrupt. Then, an "ION" instruction (interrupt on) allows interrupts to be recognized by the processor. The flow of events in a program looks something like this:

Main program

```
     .
     .
     .          set interrupt enable on devices
     .
    ION         Allow processor to recognize interrupts
     .
     .
     .
  continue      interrupt and main program run
                contemporaneously
     .
     .
     .
```

When a device interrupts, the current value of the program counter is stored in location 0 and the machine continues to execute instructions upward in memory (i.e., 1, 2, . . .). Most often the instructions look like this:

Location	Instruction	Comments
0	0	Storage for return
1	JMP I 2	Put indirect jump in location 1.
2	SERVELOC	The name of the service routine.

"JMP I 2" means to jump indirectly to the location specified in location 2. Location 2 contains the address of a routine that determines which device interrupted. In this case, SERVELOC is the label associated with the service routine. The assembler translates the name into an address. The user is responsible for putting the correct values in locations 1 and 2 before turning on the interrupt system. Thus, each time an A/D interrupt occurs (perhaps driven by a clock or an external signal), the current value of the PC (program counter; the next location in the program to be executed) is stored in location 0 and the PC reset to the address contained in location 2. In the service routine, beginning at the address contained in location 2, the following events occur:

```
SERVELOC:   enter the interrupt handler
            save the flag register (i.e., carry, etc.)
            did device1 interrupt? if yes, go to device 1
            did device2 interrupt? if yes, go to device2
 restore:   restore flags
            exit and return to main program
 device1:   take action (e.g., sample and store)
            goto restore
 device2:   take action (e.g., print character)
            goto restore
```

In the code that asks which device interrupted, all device flags are tested, e.g., in PDP-8 code:

```
ADSK          /Is it the A to D that interrupted? Skip the next
              /instruction if A/D is done. If it was the A/D
JMP .+2       /skip the next instruction and
JMP ADSER     /jump to service location for the A/D.
CLSK          /Is it the clock?
JMP .+2       /.+2 means current location (.) +2
JMP CLOCK     /jump to clock service routine
TSK           /Is it the teleprinter?
JMP .+2
```

```
JMP TELE                              .
                                      .
     .          /more devices can be added here
     .
     .
                /restore flags
JMP I 0         /return
     .
     .
     .
```

This line of code is called a skip chain. Exiting the service routine is performed by a JMP I 0, which simply means the PC is set to the location stored in location 0 (recall that we stored our return address there when the interrupt occurred).

8.2.2 PDP-11 Interrupts

The PDP-11 uses a vectored interrupt system. Rather than having each device interrupt produce a hardware jump to location 0, each device has its own vector address. This architecture eliminates the need for software polling (that is, the skip chain checking of flags of each device). For example, the clock (e.g., KW11-P) has an interrupt vector at address 0104. Two sequential locations starting at 0104 are used: 0104 should contain the address of the service routine for the clock and 0106 the value that is to be put in the processor status word (PSW) during the time the interrupt is serviced. At the time of the interrupt (e.g., when the clock ticks, the bell sounds, etc.), the value of the PC and the PSW are pushed on the stack and the PC and PSW in the vector locations (e.g., 104 and 106) are loaded into the PC and the PSW. On return from an interrupt (RTI), these two values on the stack are reloaded into the PC and PSW and the SP is incremented to where it was before. The automatic pushing of information onto the stack is the same as if the instructions

```
mfps -(sp)
mov pc, -(sp)
```

had been executed. "mfps" and "mtps" are two instructions used to move a byte from or to the PSW. Figure 8.1 shows the state of the machine just before and after an interrupt occurs. Clearly, the PSW may contain different values in the z, n, c, v bits depending on what the program is doing when the interrupt occurs. Therefore, it is essential to store the PSW as well as the PC.

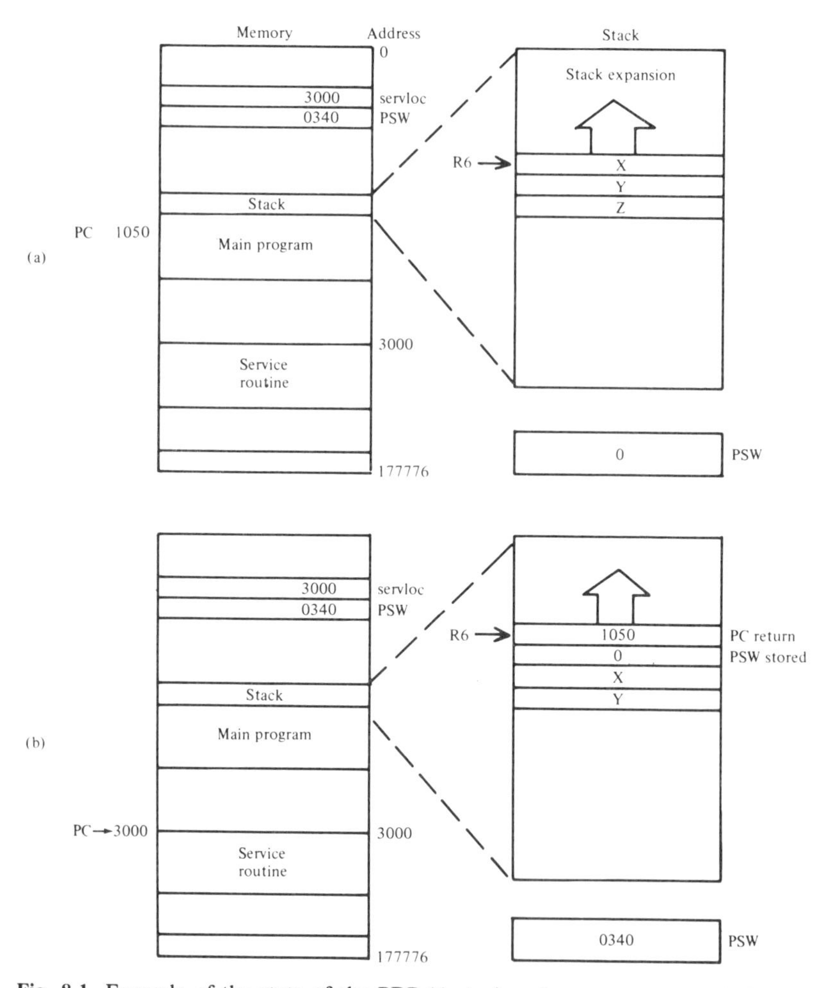

Fig. 8.1. Example of the state of the PDP-11 stack and program counter (PC) (a) just before and (b) after an interrupt. The PC is initially 1050 and the service routine is loaded in memory at 3000 in this example.

Table 8.1 shows a memory map for an 11/03. Examples of typical device locations and interrupt vector locations are given in this table. One must load the interrupt vector locations for all devices being used before setting bit 6 in the CSR of devices that will be allowed to interrupt. For example, the two 'as' instructions:

```
mov $clkser, *$0104
mov  $0340, *$0106
```

TABLE 8.1

AN EXAMPLE MEMORY MAP FOR AN 11/03

Location	Description
000	Reserved—beginning of memory
004	Bus time out, illegal instructions
010	Illegal and reserved instruction
014	BPT trap
020	IOT trap
024	Power fail trap
030	EMT trap
034	TRAP trap
060	Keyboard interrupt address
062	Keyboard PSW
064	Teleprinter interrupt address
066	Teleprinter PSW
104	Clock KW11p
106	Clock PSW
130	A/D
132	A/D PSW
300	Floating vectors:
	DI/O
	Other serial lines
376	Last location for interrupt vectors
400-157777	Reserved for programs
167760	DI/O CSR
167762	DI/O output buffer
167764	DI/O input buffer
177560	Keyboard reader CSR
177562	Buffer
177564	Teleprinter CSR
177566	Teleprinter buffer
176760	D/A #1
176762	D/A #2
176770	A/D status
176772	A/D buffer
177777	END OF MEMORY

fill the PC and PSW locations for the clock. The value placed in the PSW location (106) is the priority to be in effect during the execution of the service routine. "clkser" is the entry location to the clock service routine.

Figure 8.2 shows the placement of bits in the PSW. Note that an octal '7' (= binary 111) when placed in bits 5, 6, and 7 (the priority bits) produces 0340_8. A priority of 7 means that nothing else can interrupt until

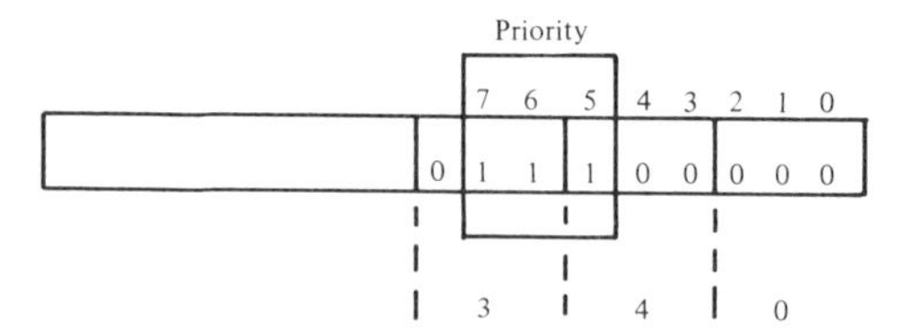

Fig. 8.2. Location of priority bits in the PSW.

control is relinquished (i.e., the interrupt routine terminates). On the 11/03, only two interrupt levels are available, 0 and 4. The device physically nearest the CPU board has the highest priority. There are no assignable vertical priority levels (BR7-4) as on larger 11s; only horizontal priority is available (see Fig. 3.2). The clock is often placed nearest the LSI-11 and memory card since the clock is usually involved in real-time events. On larger machines (e.g., 11/34, 11/70) all priority levels 0–7 are present but, as noted in Chapter 3, five levels for bus requests (NPR, BR7–BR4) are for hardware, while levels BR3–BR0 can be used for software. If two devices are assigned different priority levels (e.g., 4 and 5) and each request servicing simultaneously, the one with the highest priority will be serviced first. Since we use the stack to store the current PC and PSW, multiple levels of interrupts are easy to handle. When a device interrupts, the current PC and PSW are pushed onto the stack. When the service routine terminates, the program can resume execution wherever it was when it was interrupted.

The 'C' analog to the machine code for loading the PSW and PC locations in the interrupt vectors is

```
CLK->intvec = &clkser;
CLK->psw    = 0340;
```

Note that we have employed the common structured method of specifying two adjacent locations, i.e.,

```
struct { int intvec,
             psw;
       };
# define CLK 0104
```

where "&clkser" specifies the address of the clock routine used for servicing. Often this address is defined globally so that the service routine can be in a separate file.

The typical 11/03 setup routine is

Set up vector;	Each device has its own vector.
Set up PSW;	On PDP 11/03, PSW = 0200
Enable devices;	Set device registers of individual devices.

.
.
.

and in the interrupt routine:

```
test for error;
do whatever needs doing;
return;
```

As described earlier, the 400_8 locations between 0 and 400_8 contain a variety of device interrupt locations. Each device that is to be used in interrupt mode must be set. Moreover, it is prudent to set all the other vectors to avoid problems with spurious interrupts. Some 11/03 systems occasionally produce random interrupts with no apparent cause; that is, a jump to a location in the memory between 0–400_8 occurs for no clear reason. This problem has been observed in a number of 11/03-based computer systems. The remedy is straightforward. If zeros are placed in all unused memory locations from 2 to 400 and the value 2 (the code for rti) is placed in location 0, any spurious interrupt will simply produce an rti. If programs are written in 'as' or Macro, the interrupt vectors and the rti code can easily be added. If 'C' programs are used with interrupts, .s versions of the code can be edited to add the rti code. Another strategy can be used by accessing a routine called "bouncer." "bouncer" can be written in 'C' for use in returning from spurious interrupts.

```
bouncer( ) {
        rti( ); };
```

The address of bouncer is placed in all alternating even locations between 0 and 400 (i.e., 0, 4, 10, 14, . . .) before the device registers are filled. That is,

```
for (i = 0, i<0400; i=+ 4) *i = &bouncer;
```

The reader should note that a routine called "rti( )" is called in bouncer. rti( ) is an "as" program named _rti containing

```
rti = 2
rti
```

Recall that the _ name construction in 'as' permits calling a routine in 'C' by name( ). The same result could also have been achieved by editing a .s version of bouncer without the rti( ) call to include the 'as' code.

8.3 TERMINAL EXAMPLE

The method for programming interrupts for the terminal is typical of other serial devices (e.g., the paper tape reader). For the console DLV-11 card the interrupt vector is at location 060. Figure 8.3 shows the definitions used in an example of terminal operation. Terminal I/O registers and the relevant bits in the CSRs are defined. KLDONE (ttydone) is 0200 for bit 7 and INTON (interrupt on) is 0100 for bit 6.

Two main routines are used: tty.c and ttyintr.c are shown in Figs. 8.4 and 8.5. The definitions in tty.h are included at the outset. In main, the terminal is set up by calling "ttysetup( )." This routine turns off the printer interrupt and sets up the interrupt vector using the callC routine (described below). The PSW in the interrupt vector is set to 340 by passing KLPSW via callC to location 62. Then, after turning on the interrupt for the reader (keyboard), the routine terminates and returns to main.

In main, a flag ("ttyready") is tested. "ttyready" is set in ttyintr whenever a complete line is typed in. Basically, this example echoes characters as they are typed in and then prints the line over when a carriage return is typed. Whenever a key is struck, the program vectors via location 60 and callC to ttyintr. A character is read and put in charbuf. Then the character is printed. If a \r is found, a \n is inserted and the ttyready flag is set to indicate to main that the character buffer can be printed out. After printing, the flag is lowered (set to zero).

Figure 8.6 shows the .s file corresponding to the ttyintr.c file. For an interrupt service routine to function properly, a rti must be inserted at the end of the routine. Since ttyintr.c is originally typed in as a 'C' subroutine, the compiler knows nothing about the user's intention to make it an interrupt service routine. Consequently, the user must modify the routine. Two or more simple modification strategies exist.

One method is to use an 'as' program to handle the interrupts. Figure 8.7 shows a routine that will set up as many as 10 interrupt vectors. The routine "callC" determines the interrupt service routine addresses, saves r0, r1, and provides an rti at the end of the interrupt. Registers r2, r3, and r4 are saved by csv.

```
/*************************************************************************
*       tty.h - file of definitions for use in tty.c program
        in Figure 8.4.
*       To include this file in another file:
*       #
*       #include "tty.h"
*       where tty.h is in the current directory
**************************************************************************/

#define MASK 0177
#define KLINTV 060
#define KLREAD 0177560
#define KLPRINT 0177564
#define KLPSW 0340
#define KLDONE 0200
#define INTON 0100


*************************************************************************/

*               Structure definitions:

                        tps is the teleprinter status
                        tpb is the teleprinter buffer
                        intvec is the interrupt vector location
                        psw is the processor status word location
                        tkb is the keyboard buffer
                        tks is the keyboard status


*********************************************************************/

truct {
   int      tps,
            tpb;
;                       /* printer status and buffer */
truct {
   int      intvec,
            psw;
;                       /* interrupt vector and processor status word */
truct {
   int      tks,
            tkb;
;                       /* keyboard status and  buffer */
```

Fig. 8.3. Definitions file. Used in example program in Fig. 8.4.

```
#

#include "/usr/john/text/fig8/fig8.3"

extern char charbuf[100];             /* use extern when routines are in
                                         different files */
                                      /* and common variables are to be
                                         accessed */
extern int  ttyready,
            i;

extern  ttyintr ();

main () {
    int     j;
    ttysetup ();
    while (1) {
        if (ttyready != 0) {
            for (j = 0; j < i; j++) {
                while (!(KLPRINT -> tps & KLDONE));
                KLPRINT -> tpb = charbuf[j];
            }
            ttyready = 0;
            i = 0;
        }
    }
}

ttysetup () {

    extern int  ttyready,
                i;
    KLPRINT -> tps = 0;
    callC (KLINTV, &ttyintr, KLPSW); /* set up the interrupt service
                                         routine */
    ttyready = i = 0;
    KLREAD -> tps = INTON;
}
```

Fig. 8.4. "tty.c". This is a test program for typing a string of characters on the terminal and echoing them whenever a carriage return is typed. The interrupt routine "ttyintr( )" shown in Fig. 8.5 is used to service the interrupt.

For example,

.

.

main program

.

callC (KLINTV, &ttyintr, KLPSW)

.

.

```
#

#include "/usr/john/text/fig8/fig8.3"

ttyintr () {
    extern char charbuf[100];
    extern int  ttyready,
                i;
    charbuf[i] = KLREAD -> tkb & MASK;
    KLPRINT -> tpb = charbuf[i];
    if (charbuf[i++] == '\r') {
        charbuf[i++] = '\n';
        ttyready++;
    }
}
```

Fig. 8.5. "ttyintr.c" An interrupt routine for the tty.c program. Note that the callC(. . .) function included in tty.c will handle the register saves and supply the rti for the routine. If callC(. . .) or equivalent is not used, the user must edit the .s code to supply register saves and the rti.

```
.text
.globl  _ttyintr
_ttyintr:
~~ttyintr:
jsr     r5,csv
mov     _i,r0
mov     *$-216,r1
bic     $-200,r1
movb    r1,_charbuf(r0)
mov     _i,r0
movb    _charbuf(r0),r0
mov     r0,*$-212
mov     _i,r0
inc     _i
cmpb    $15,_charbuf(r0)
jne     L2
mov     _i,r0
movb    $12,_charbuf(r0)
inc     _i
inc     _ttyready
L2:L1:jmp       cret
.globl
.data
```

Fig. 8.6. A ".s" equivalent of ttyintr.c. No register saves or rti code is supplied here.

```
/       callC.as

/       Routines to call a C routine from an interrupt.
/
        wait    =       1
        rti     =       2
        mtps    =       106400^clr
        mfps    =       106700^clr
/
/       To set up the vector and internal tables use the
/       following routine:
/
/       callC(vector, &isr, newps);
/
/       vector  : vector address  (e.g. 060 for tty)
/       &isr    : address of C interrupt service routine
/       newps   : will become 2nd word of int vector
/
/       returns -1 on error [ table full ]
/
/       The interrupt service routine will be called
/       as follows:
/
/       isr(r1, nps, r0, pc, ps)
/
/       r1      : saved r1
/       nps     : ps from vector [ useful in trap handler etc. ]
/       r0      : saved r0 [ trap handler etc. ]
/       pc      : saved pc
/       ps      : saved ps
/
/         This routine is NOT called by value. The saved
/       registers will be copied back to the previous routine.
/
/       if you don't care about the args, just ignore  them --
/
/       isr()
/
        .globl  _callC, csv, cret
        .text
/
_callC: jsr     r5, csv
        mov     eptr, r1                / get address
                                        / of next free table entry
        cmp     r1, $eend               / table full?
        blo     1f                      / no -
        mov     $-1, r0                 / yes - return error code
        jmp     cret

1:      mov     4(r5), r2               / get vector address
        mov     r1, (r2)+               / set up vector
        mov     8.(r5), (r2)            / new ps
        mov     6(r5), 4(r1)            / save isr address in table
        add     $6, r1                  / update free entry pointer
        mov     r1, eptr
        clr     r0                      / success code
```

Fig. 8.7. The callC(. . .) routine for directing interrupts to service routines written in 'C'.

```
        jmp     cret
/
/       interface code to C
/
intr:   mfps    -(sp)                   / get nps
        bic     $!377, (sp)             / 8-bits only
        mov     r1, -(sp)               / save r1
        jsr     pc, *(r0)               / call user isr
        mov     (sp)+, r1               / restore r1
        tst     (sp)+                   / pop off nps
        mov     (sp)+, r0               / restore r0
        rti                             / return
/
/
/       The address of the user isr is stored after "entry".
/       When an interrupt occurs  these are the routines executed
/       directly from the vector.  They  call
/       the user isr in a rather round about way. Note that
/       there is NO return from the jsr r0's.

/
        .data
entry:  jsr     r0, intr; 0             / isr address will replace 0
        jsr     r0, intr; 0
        jsr     r0, intr; 0
        jsr     r0, intr; 0
        jsr     r0, intr; 0
        jsr     r0, intr; 0
        jsr     r0, intr; 0
        jsr     r0, intr; 0
        jsr     r0, intr; 0
        jsr     r0, intr; 0
        jsr     r0, intr; 0
        jsr     r0, intr; 0
        jsr     r0, intr; 0
        jsr     r0, intr; 0
        jsr     r0, intr; 0
        jsr     r0, intr; 0
eend    = .

eptr:   entry                           / free entry pointer
/
/-------------------------------------------------------------
/       wait();         wait for an interrupt to occur
/
        .globl  _wait
        .text

_wait:  mfps    -(sp)
        mtps    $0
        wait
        mtps    (sp)+
        rts     pc
```

Fig. 8.7. (continued)

In this example, callC passes the interrupt location KLINTV, the address of ttyintr and KLPSW, the processor status value to be operative during the interrupt.

'callC' will return a −1 if the number of interrupt locations set up is greater than 10. To check for this condition, it is prudent to include the following code:

```
if (callC(KLINTV,&ttyintr,KLPSW) == -1)print(''callC failed\n\r'');
```

An 'as' call for wait is included in Fig. 8.7. This function can be accessed in 'C' by

```
wait( );
```

Another method for handling the interrupts is to edit the .s version of the 'C' interrupt routine and insert a rti at the end of the code. Also, rti = 2 must be defined for the assembler at the beginning of the .s file. The registers employed in the routine must be saved at the outset and restored prior to returning. For example, ttyintr.s uses r1 and r0. These registers might be used in the main program that was interrupted and thus must be saved during the interrupt. To accomplish this operation, two variables, reg0 and reg1, could be specified as integers in the 'C' program that would then become _reg0 and _reg1 in the .s file. The statements

```
mov r0, _reg0
mov r1, _reg1
```

and

```
mov _reg0, r0
mov _reg1, r1
rti
```

would be inserted at the beginning and end of the .s file. Note that the register saves should appear before the "jsr r5, csv" so that r0 can be preserved before it is used in csv. The compilation strategy would be

```
% cc -S ttyintr.c
% [edit ttyintr.s and insert "rti = 2" (definition),
  "rti" (command), and register store definitions]
% as - ttyintr.s
% mv a.out ttyintr.o
% cc -2 tty.c ttyintr.o -lc
% mv a.out runfile
```

"runfile" can now be executed on the 11/03.

The 'C' library supplies the csv and cret routines in the compilation. The "−2" option adds a 1000_8 location header to the beginning of the code (see Chapter 5).

From the above discussion, it should be apparent that either the use of callC( ) or the file-editing strategy will work. The former, however, is clearer and simpler to implement. The latter was presented simply to indicate that alternative methods do exist.

8.4 A/D INTERRUPT EXAMPLE IN 'as'

The following example uses only one device, the A/D, as the interrupt driven device. We shall employ the facility of driving the A/D from an external source, e.g., a signal generator. Typically, most A/Ds allow external triggering by setting bits in the CSR (see Chapter 7 for A/D status register configuration). An external source, producing a square wave or pulse, is connected to a line designated by the manufacturer for external input and set to some frequency. The basic idea is that on a rising or falling edge (also often selectable) of an input pulse, the A/D will be started. Consider the example 'as' program in Fig. 8.8. The A/D vector address is contained in location 130. When the A/D completes taking a sample, the interrupt occurs and the A/D service routine is initiated via an indirect jump through location 130.

First, typical definitions are made. An offset of 1000 bytes (.=.+1000) is added to the beginning of this program so that it will start at location 1000 (base 8). This step may be omitted if the loader adds the value in automatically (e.g., in cc; the −2 option). At label 'start', the current value of the PC is put into R6 (the SP) and the stack pointer is decremented in the next instruction. After these two operations, the SP points to the address of start. This manipulation is carried out to set up the stack immediately below (closer to 0) the program. The next major problem is to find the address of adser, the A/D service routine. Since a program is not always run in the same location in memory, it is often useful to compute the location of the service routine in the program itself, rather than defining it prior to compiling the program. At x:, we find out where we are now in memory, then add the number of locations to the PC between where we are now (the PC address) and the service routine. This value is moved to location 130 to set up the vector. After 0200 is moved to address 132, we are ready to turn on the interrupts. Note the use of $102 for enabling the interrupts. The 1 is used for setting bit 6. In this example, bit 1 (2_8) must be

```
        STATUS = 176770          /A/D status
        DATA   = 176772          /A/D data
        VECTOR = 130             / A/D vector
        PSW    = VECTOR +2       / Processor status
        rti = 2

.=.+1000                         / offset for use with RT-11

start:  mov pc,sp
        tst -(sp)
        mov $0, alldone          / set alldone flag to zero
        mov $buf, r1
        mov $1000,count
x:      mov pc, r0               / r0 = addr((x)+2
        add $adser-x-2, r0       / add offset
        mov r0,*$130             / or mov $adser *$130
        mov $200, *$132          / fill PSW
        mov $0102, STATUS        / or *$176770
                                 / 102 = int enable + external enable, ch (

        jsr pc, display

adser:  tst STATUS               / or *$0176770
        bmi error
        mov *$DATA,(r1)+         / move data sampled into list
        dec count
        beq quit
        rti

quit:   mov $0, STATUS           / turn off A/D
        bis $1, alldone          / set done flag
        rti

done:   sys exit                 / exit when done sampling
                                 / exit as shown in figure 8.11 using an
                                 / emt if you are using rt-11
                                 / the sys exit is used with MiniUNIX
alldone:        0                / flag to set when sampling done
error:                           / print an error message
count:  0
ptr:    0
buf:    .=.+1000                 / setup a buffer for the data
display:                         / any conventional display here
                                 / Do not use r1 unless you save it.
                .                / Include a test in each pass
                .                / through the display loop
        tst alldone              / to see if all samples are taken
        bne done                 / if so, exit.
                .
                .
```

Fig. 8.8. Outline of a sampling program using an externally driven A/D.

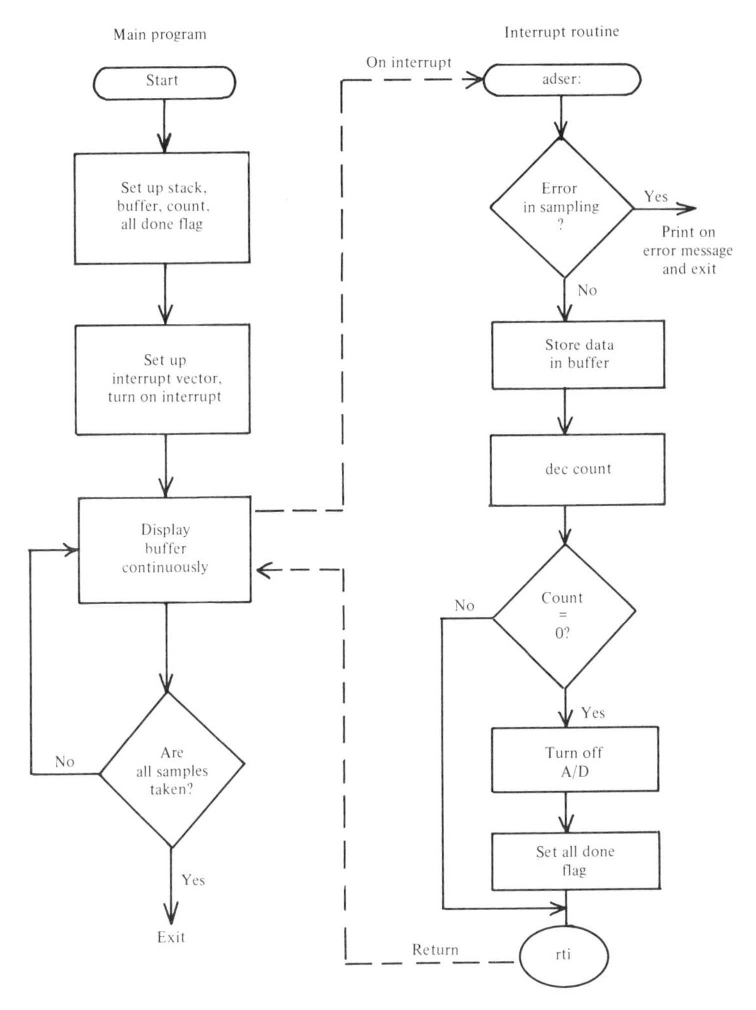

Fig. 8.9. Flow chart of sampling program outlined in Fig. 8.8.

set to allow an external triggering. Figure 8.9 shows a flow chart of the operation of this program.

8.5 CLOCK AND A/D INTERRUPT EXAMPLE IN 'as'

Figure 8.10 shows an example of using the clock as the interrupting device. In brief, the clock is set up to tick at a prespecified rate. At each

```
          STATUS = 0172540
          BUFFER = 0172542
          COUNTER= 0172544
          ADSTAT = 0176770
          CLOCKVEC = 0104
          emt = 104000              / the code for the emulator trap
          rti = 2


begin:    mov $buf, r1
          mov $clkser, *$CLOCKVEC
          mov $200, *$CLOCKVEC+2
          mov $111, *$STATUS        / set clock for down count,
                                    / 100khz,start,enable interrupt
display:                            / insert any display here
/         .
/         .
/         .


clkser: mov $01,ADSTAT              / start A/D
loop:     tstb *$ADSTAT             / is it done?
          bpl loop                  / no , branch back
          mov *$ADSTAT+2,(r1)+      / store data away
          rti


exit1:    emt ¦ 350                 / return to RT-11
buf:      .=.+1000                  / a place to store the data
```

Fig. 8.10. 'as' program outline for setting up the clock to sample the A/D. Note that r1 must be saved in clkser routine if it is used in the display routine. We assume that the stack is already set up; if not, r6 must be set to someplace in memory. Exits may be to RT-11, the MINIUNIX system, or simply a halt.

tick, the A/D is started. Note carefully the distinction between this example and the previous example, i.e., the interrupt vector is now 104—the clock's. It is not necessary to use 130 at all.

An important point is that we loaded 0104 with the address of where clkser will be loaded and, therefore, did not have to compute that address. Note that this direct use of an address works only when you know where loading of the program occurs. The value moved to the CSR is $111—or enable interrupt + downcount + repeat + 100 kHz + start. There are a variety of differences between this example and the previous one as shown in the code.

8.6 CLOCK, A/D EXAMPLE IN 'C'

Figure 8.11 shows a routine in 'C' that does essentially the same thing as the previous example. A buffer has been declared to start at STARTADD. STARTADD is defined to be 400 in ad.h (Fig. 8.14)—it could have

```
include "/usr/john/ttytest/tty.h"
include "/usr/john/text/fig8/ad.h"

xtern  clkintr ();
xtern char *adpoint STARTADD;

*************************************************************************

      main() calls startsam, fills memory with data and then
      calls display() to display the sampled data.

*************************************************************************/

ain () {

   int     done;
   startsam ();                     /* start sampling */
   while (1) {

       if (adpoint >= ENDADD)
           break;
   }
   CLOCK -> clkstbuf = CLOCKSTOP;
   while (1) {
       display ();                  /* display forever.  Insert a keyboard
                                       check to return to monitor , if
                                       desired */

   }

*************************************************************************

      startsam() starts the clock and returns

*************************************************************************/
```

Fig. 8.11. Example program in 'C' for sampling the A/D. A function named transfer( ) is included for interprocessor data transfer (see Appendix C).

```
startsam () {

    callC (CLINTV, &clkintr, CLPSW);
    CLOCK -> clkpreset = TICKS;
    CLOCK -> clkstbuf = CLOCKSTART;

}

/***********************************************************************

        display() first transfers the sampled data to another
        PDP - 11 and then displays the sampled data in 500 sample
        chunks.

***********************************************************************/

display () {
    int     NBYTES;
    char    *outfile;

    int     j;
    int     i;
    extern char *adpoint;
    adpoint = STARTADD;
    NBYTES = ENDADD - STARTADD;
    outfile = "test";
    trans (adpoint, outfile, NBYTES);
    printf ("now display \r\n");
    while (1) {
        for (i = -2000; i <= 2000; i =+ 8) {
            for (j = 0; j < 512; j++)
                continue;           /* wait briefly */
            DA -> da1 = *adpoint++;
            DA -> da2 = i;
        }
        if (adpoint >= ENDADD)
            break;
    }
    return;
}
```

Fig. 8.11. (continued)

been anywhere, but this location was convenient since one version of the program was originally loaded at 060000 with the MINIUNIX* system. If RT-11 was used to load the program at 1000, then the buffer would be placed above the program in memory by redefining STARTADD and END-ADD. "adpoint" was defined to point at characters and A/D samples were truncated to 8 bits before storing in memory. First, startsam( ) is called; this routine sets up and starts the clock. Next, in the while loop,

* MINIUNIX is a trademark of Bell Laboratories.

memory is filled with sampled data from STARTADD to ENDADD. Then, the clock is stopped and the data displayed. Of particular interest is the routine in clkintr.c (Fig. 8.12) which samples 4 channels:

(a) The A/D is started: AD−>adstbuf = buffer = 01

(b) The value buffer is set to ADSTART = 01 initially and 400 is added on each loop around the "for" loop to yield:

Pass	A/D CSR
1	01
2	401
3	1001
4	1401

The addition is accomplished by use of the statement: "buffer= + MUX". Examine what happens when the above numbers are put in the A/D CSR. You will find that a different channel is selected on each pass!

(c) As usual, we wait for the ADDONE flag to come up and store the value away in memory. The construction

*adpoint = (AD−>in >>4)

first shifts (>>) the value in the A/D buffer register four places right, i.e., makes it byte size since it was 12 bits initially (12 bits − 4 bits = 8 bits = 1

```
#

int     buffer,
        i;
#include "/usr/john/text/fig8/ad.h"
#define NUMSHIFT 4                  /* number of shifts */
#define NUMCHAN 4                   /* number of A/D channels */

extern char *adpoint;


clkintr () {
    buffer = ADSTART;               /* buffer is the value
                                     loaded into the a/d buffer */

    for (i = 0; i < NUMCHAN; i++) {

        AD -> adstbuf = buffer;
        buffer =+ MUX;
        while (!(AD -> adstbuf & ADDONE));
        *adpoint = (AD -> adin >> NUMSHIFT);
        DA -> da1 = *adpoint++;
    }
}
```

Fig. 8.12. clkintr.c. The interrupt routine called by example program in Fig. 8.11.

byte). Then, the sampled and shifted value is stored in memory—in bytes. The samples in memory appear as shown in Fig. 8.13.

Unless one wishes to simply observe that data are, in fact, being sampled, data points in each channel will need to be shifted vertically on the scope face. Without shifting, the values displayed from the four channels will be intermingled. One possibility for producing offsets is to define initially an array of values corresponding to dc levels that can be added to each sample prior to display. For example, defining

int offset [4] {−1000, −512, 512, 1000};

would initialize offset[ ] to −2.5, −1.25, 1.25, and 2.5 V. Then, modification of the display statement as

DA−>da1 = *adpoint++ + offset[i];

will add the specified DC level to each sampled value as it is displayed.

The display routine in Fig. 8.12 incorporates a call to the function "trans" for moving sampled data to another computer. The conceptual base for this operation is that a sampling program can be used on an 11/03 and data moved to a larger machine for analysis. The function used here is

trans(adpoint, outfile, NBYTES);

where

"adpoint" points to the data
"outfile" points to the name of the file
to be created

Sample 2	ch 1	Sample 1	ch 0	400
Sample 4	ch 3	Sample 3	ch 2	402
Sample 6	ch 1	Sample 5	ch 0	404
Sample 8	ch 3	Sample 7	ch 2	406
				410
o o o		o o o		

Fig. 8.13. Storage arrangement of 8-bit samples in memory as used in the routines shown in Figs. 8.11 and 8.12.

```
#define TICKS 1562
#define STARTADD 0400
                /* STARTADD is the address where we start storing data */
                /* ENDADD is the address where we stop storing data */
                /* These values will write over MiniUNIX in this example */
#define ENDADD 040000
                /* For RT-11, start storing data above program */
                /* for example, at 02000        */
#define MUX 0400
#define ADSTART 01
#define CLOCKSTOP 00
#define MASK 0377
#define CLINTV 0104
#define CLPSW 0340
#define CLOCKSTART 0111
#define DA 0176760
#define AD 0176770
#define CLOCK 0172540
#define ADDONE 0200

/**********************************************************************

        structure definitions:

                intvec,psw : the two word interrupt and psw pair
                clkstbuf
                clkpreset
                clkcount :   three words allocated for the real time clock
                da1,da2:        the D/A
                adstbuf,adin:    the A/D

**********************************************************************/


struct {
    int     intvec,
            psw;
};

struct {
    int     clkstbuf,
            clkpreset,
            clkcount;
};

struct {
    int     da1,
            da2;
};

struct {
    int     adstbuf,
            adin;
};
```

Fig. 8.14. Definitions file for program listings in Figs. 8.11 and 8.12.

```
.data
.globl  _adpoint
_adpoint:
400
.text
.globl  _main
_main:
~~main:
isr     r5,csv
~done=177770
tst     -(sp)
jsr     pc,_startsam
L2:tst  $1
jeq     L3
cmp     $40000,_adpoint
jlos    L3
jbr     L2
L3:clr  *$-5240
L4:tst  $1
jeq     L5
jsr     pc,_display
jbr     L4
L5:L1:jmp       cret
.text
.globl  _startsam
 startsam:
~~startsam:
jsr     r5,csv
mov     $340,(sp)
mov     $_clkintr,-(sp)
mov     $104,-(sp)
jsr     pc,*$_callC
cmp     (sp)+,(sp)+
mov     $3032,*$-5236
mov     $111,*$-5240
L6:jmp  cret
.text
.globl  _display
_display:
~~display:
jsr     r5,csv
~outfile=177766
~i=177762
~j=177764
~NBYTES=177770
sub     $10,sp
mov     $400,_adpoint
mov     $37400,-10(r5)
mov     $L8,-12(r5)
mov     -10(r5),(sp)
mov     -12(r5),-(sp)
mov     _adpoint,-(sp)
jsr     pc,*$ trans
cmp     (sp)+,(sp)+
mov     $L9,(sp)
jsr     pc,*$_printf
L10:tst $1
jeq     L11
mov     $-3720,-16(r5)
L12:cmp $3720,-16(r5)
jlt     L13
clr     -14(r5)
L15:cmp $1000,-14(r5)
jle     L16
jbr     L17
L17:inc -14(r5)
jbr     L15
L16:movb        *_adpoint,r0
mov     r0,*$-1020
inc     _adpoint
mov     -16(r5),*$-1016
L14:add $10,-16(r5)
jbr     L12
L13:cmp $40000,_adpoint
jlos    L11
jbr     L10
L11:jbr L7
L7:jmp  cret
.globl
.data
L8:.byte 164,145,163,164,0
L9:.byte 156,157,167,40,144,151
   .byte 163,160,154,141,171,40,15,12,
```

Fig. 8.15. 'as' version (.s) of program in Fig. 8.11. Note construction of function calls.

and

"NBYTES" specifies the number of bytes
to be transmitted.

Various implementations of this function are possible including both serial and parallel transmission of data. The discussion of programs for these types of communication is outside the scope of this text. However, a program for accomplishing such communication is listed in Appendix C.

In the display loop, "adpoint" is moved across the data displaying values from the buffer. At the end of the buffer, the routine is terminated. Various definitions used in this sampling program are shown Fig. 8.14 including the structure definitions employed.

Figure 8.15 displays the .s file rendition of the .c file for sampling. The reader will find it instructive to make a detailed comparison of the .c and .s versions of the file. The .s code for clkintr is shown in Fig. 8.16.

```
.comm   _buffer,2
.comm   _i,2
.comm   _adpoint,2
.text
.globl  _clkintr
_clkintr:
~~clkintr:
jsr     r5,csv
mov     $1,_buffer
clr     _i
L2:cmp  $4,_i
jle     L3
mov     _buffer,*$-1010
add     $400,_buffer
L5:bit  $200,*$-1010
jne     L6
jbr     L5
L6:mov  *$-1006,r0
ash     $-4,r0
movb    r0,*_adpoint
movb    *_adpoint,r0
mov     r0,*$-1020
inc     _adpoint
L4:inc  _i
jbr     L2
L3:L1:jmp       cret
.globl
.data
```

Fig. 8.16. 'as' version (.s) of program in Fig. 8.12.

8.6.1 Displays

This section describes three modifications to the display method described in the previous example. The first two methods use queue-based techniques, while the third method uses a shifting window to display a long data buffer. In this context, a queue is nothing more than a fixed length, contiguous set of memory locations.

Figure 8.17 illustrates how a buffer can be constructed for display. At the top of the figure is shown a queue in which sampled data are continuously added to the first location in the queue. Between each sample, the entire set of data is shifted right to make room for the next sample of data. In contrast, data can be added continuously to memory as shown at the bottom of the figure. When the last slot in the queue is filled, the data "wraparound" in the queue by storing the next point at the beginning of the queue as shown. Thus, data circulate round and round the queue. The method of continuously shifting the values in the buffer is cumbersome and time consuming. To contrast the two methods, we shall first consider how to write the program for physically shifting data in the queue and then use another and more efficient method employing pointers with no physical shifting of data.

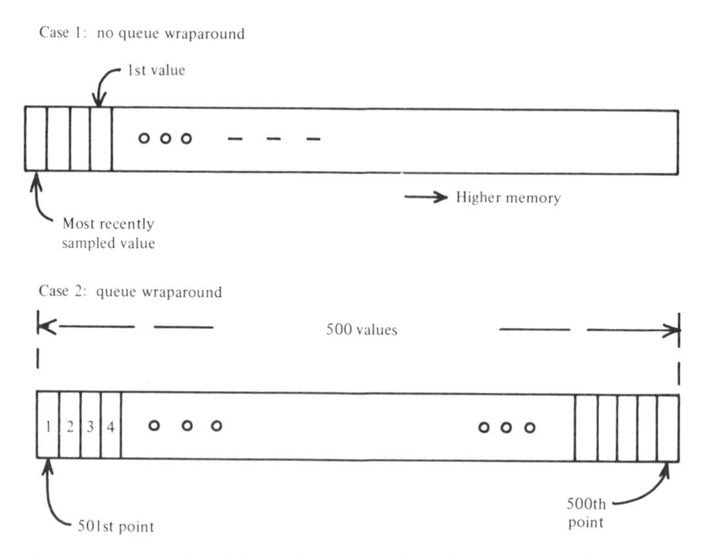

Fig. 8.17. Two uses of a fixed length buffer for data display. In case 1, the most recently sampled value is added at the beginning of the buffer as data are shifted over. In case 2, data are sequentially added to the buffer and "wraparound" after the last slot in the buffer is filled.

PERCOLATING QUEUE

Figure 8.18 displays a skeletal outline of a percolating queue sampler and display. The word "percolate" refers to the continuous shifting of data in the queue as new points are sampled. Whenever a sample is taken,

```
include "alldefs"              /* include a file here with the definitions */

nt      buf[500],
        sampleflag,
        temp;

ain () {
   startsam ();                /*   same as in fig8.11   */
                               /* display forever,add escape */
   while (1) {
       display ();
   }

isplay () {
   int      i,
            j;
   while (1) {
       for (i = 0; i < 500; i++) {
           DA -> da1 = buf[i];
           DA -> da2 = i;
       }
       if (sampleflag) {
           for (j = 498; j >= 0; j--)
               buf[j + 1] = buf[j];
           buf[0] = temp;
           sampleflag = 0;
       }
   }


kintr () {
   sampleflag++;
   AD -> adstbuf = START;
   while (!(AD -> adstbuff & ADDONE));
   temp = AD -> in;
```

Fig. 8.18. Modification of Fig. 8.11 to allow use of a percolating queue. This algorithm will work only if the sample rate is low.

a flag is set and all values in the buffer are shifted over, i.e.,

buf[j+1] = buf[j];

and the first value in the buffer filled with the new sample

buf[0] = temp;

where temp is set in the interrupt routine:

temp = AD -> in;

Note that the display can be driven from left to right or right to left, scanning the buffer from bottom to top (newest-to-oldest value) or from top to bottom.

CIRCULATING DISPLAY

Figure 8.19 shows a program fragment using pointers for the circulating display. A pointer (displaypt) is defined to point to the current data sample. Consider the diagram in Fig. 8.20. Samples are numbered in this example. The number 1 represents the first point taken. The start of the buffer is at b, data fill to c (ENDBUF), then wrap around and start filling the buffer at buf again. Display is in the alphabetic sequence abcd or dcba. As new points are sampled, they are added to higher memory locations until the end of the buffer is reached at which time wraparound occurs and the first data taken (e.g., at sample 500) overwritten. 'datapt' points to the most recently sampled value (number '7' in this figure).

In the interrupt routine we define

ENDBUF = buf + NUMPTS; e.g., let NUMPTS = 500

```
while (1) {

     displaypt = datapt;
     for (i = 0; i < NUMPTS; i++) {
          DA -> da1 = i;
          DA -> da2 = *displaypt--;
          if (displaypt == buf)
               displaypt = ENDBUF;
     }
}
```

Fig. 8.19. Accessing a data display queue using pointers. NUMPTS is the number of points in buf[NUMPTS]. "datapt" is updated as each new sample is added.

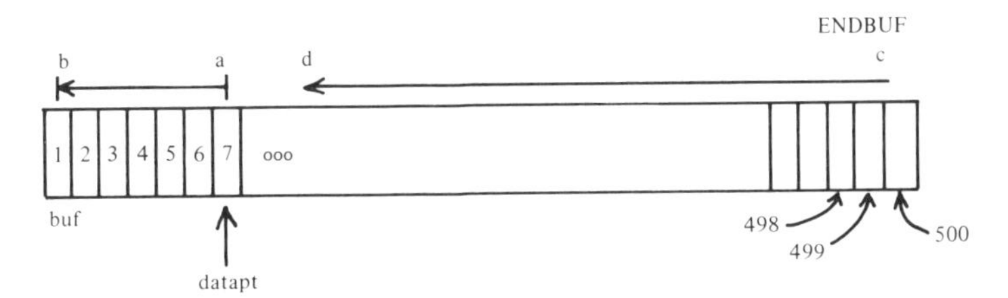

Fig. 8.20. Data acquisition and display for pointer access program in Fig. 8.19. Data are acquired from b to c. Display is abcd or dcba.

then

```
if (datapt == ENDBUF) datapt = buf;     /*reset buf*/
*datapt++ = AD -> in;                   /*store data away*/
```

The "if" statement simply resets the data pointer to the beginning of the buffer after the buffer has been filled.

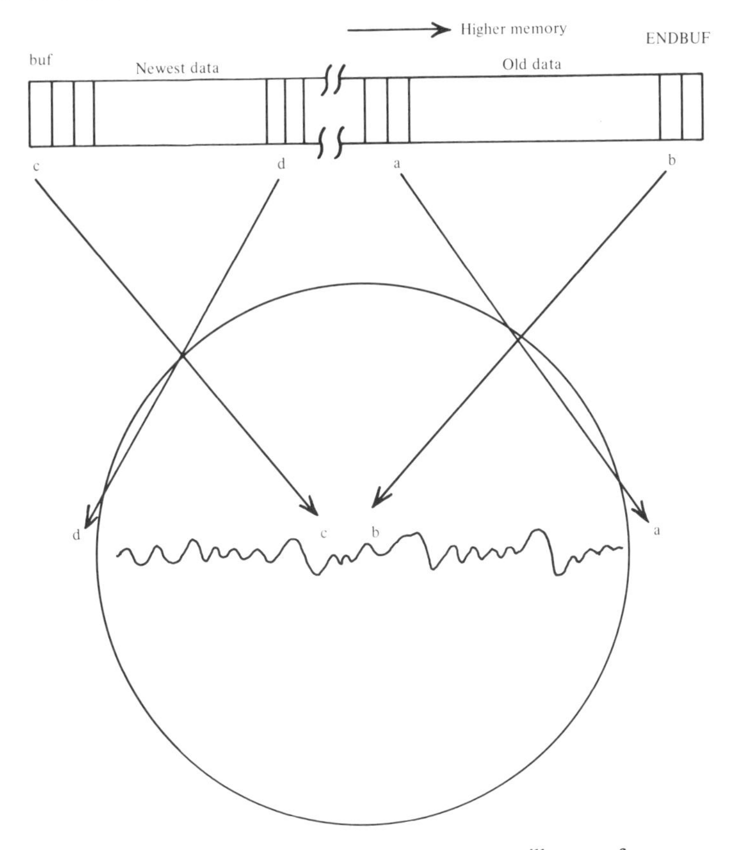

Fig. 8.21. Mapping of data from queue onto oscilloscope face.

Figure 8.21 shows diagrammatically how the sampled data are mapped onto the oscilloscope face. Data can be displayed in either the abcd or dcba sequence (i.e., either left or right on the scope face).

DISPLAY OFFSET ALGORITHM

Figure 8.22 shows an example of an algorithm used for display of data in a window moving across a continuous buffer of data. Figure 8.23 shows the memory to display map for this algorithm. The routine shown is the background display; the foreground program samples and continuously adds to buf. The *x* axis of the scope is continuously swept with i values from da2, while new values of the argument of buf are computed on each display pass. The subtraction used to compute the argument of buf requires considerable time. An alternative and faster algorithm would be to use pointers, e.g.,

```
dispt = buf + BUFSIZE;
while (1) {
        ypt = dispt;
        for (i = 0, i < 500; i++){
            DA -> da1 = *ypt--;
            DA -> da2 = i;
        }
        if (sampleflag) {dispt++;
                         sampleflag = 0;}
}
```

Now, instead of computing an address in buf 500 times, the only operations are decrementing and stuffing i into da2.

For both methods used in the display offset example, a rapid sample rate will cause problems in the display. In the example in Fig. 8.22, the buffer is offset only after the buffer is completely displayed. Thus, if the sample rate is greater than the time required to complete the "for" loop 500 times, the display will gradually get further and further behind the most currently sampled data point. For example, if the "for" loop requires 10 msec to complete display of 500 points, a sampling rate of more than 100 samples/sec would create a lagging display. One could produce a slightly different effect with high sampling rates by resetting the value of the pointer to the address of the most recently sampled data point on each update. In this case, the display would always show the 500 points prior to the most currently sampled data point. The two algorithms would produce the same effect if the sample rate was low.

```
#define BUFSIZE 499          /* define the size of the display buffer */
                             /* the size of the entire buffer for data will */
                             /* need to be defined and checked during sampling
                             */

int     offset,
        buf[5000];           /* for example, the entire data buffer (buf) can
                             */
                             /* be defined here. As samples are taken, */
                             /* check for overrun      */

display () {
    offset = 0;
    while (1) {
        k = BUFSIZE + offset;
        for (i = 0; i < 500; i++) {
            DA -> da1 = buf[k - i];
            DA -> da2 = i;
        }
        if (sampleflag) {
            offset++;
            sampleflag = 0;
        }
    }
}
```

Fig. 8.22. Display algorithm using a continuous data window.

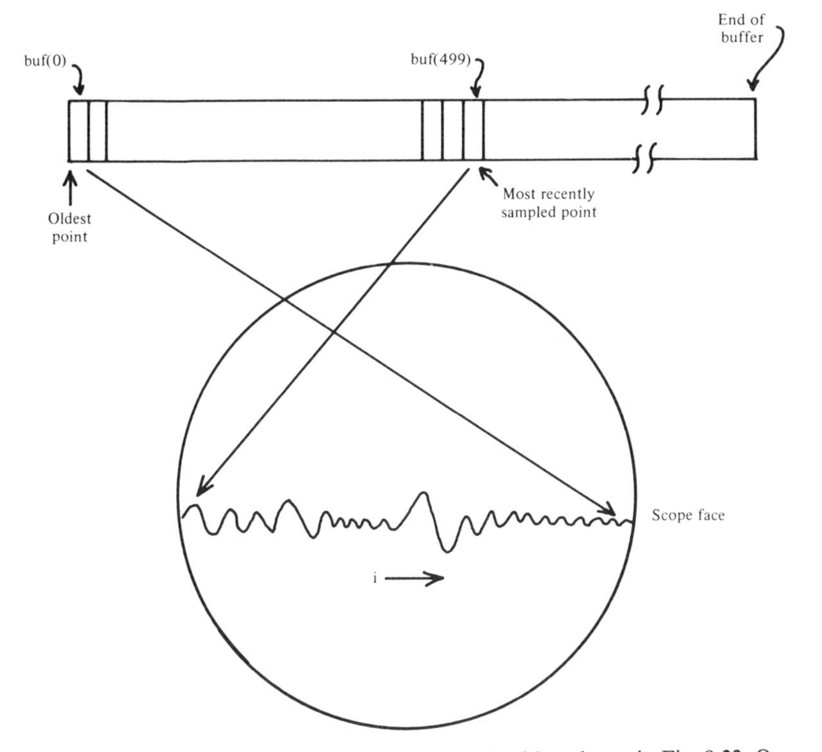

Fig. 8.23. Memory to display map for the display algorithm shown in Fig. 8.22. On pass one, display is from buf[499] to buf[0]; pass two, buf[500] to buf[1], etc.

```
/       Shown below are supporting routines for 'C' to be used on
/       an 11/03. These routines have a subtle error in them.
/       Can you tell what it is?

/       isr(); must be the first call on entry to a 'C' interrupt
/       service routine.
/       rti(); must be last call. note that this call simulates the
/       cret function and the "rti's"

        .globl _isr,_rti,_wait,_halt,_rtexit

        halt = 0
        wait = 1
        rti  = 2
        emt =  104000

        .text
_isr:   sub      $2,stkp               / save r0 and r1
        mov     r0,*stkp
        sub     $2,stkp
        mov     r1,*stkp
        rts pc

_rti:   mov     *stkp,r1        /restore r0 and r1
        add     $2,stkp
        mov     *stkp,r0
        add     $2,stkp
        tst     (sp)+
        mov     r5,r2
        mov     -(r2),r4
        mov     -(r2),r3
        mov     -(r2),r2
        mov     r5,sp
        mov     (sp)+,sp
        rti

_wait:  wait
        rts     pc

_halt:  halt

_rtexit:        emt ¦ 350       / rt-11 exit

        .data
rstk:   .=.+20  /register save stack
stkend = .
stkp:   stkend
```

Fig. 8.24. Code for Exercise 8.

The reader is encouraged to experiment with various types of displays. Most methods will yield satisfactory results for a small number of points. More care is necessary if many points are to be displayed.

EXERCISES

1. Describe a specific problem you can solve using an interrupt driven program (for examples, see problems in Chapter 9). List the peripheral devices required for the problem, their interrupt locations and the register values you will use on an 11/03 (i.e., specify where cards are to be located in memory). Second, repeat the exercise assuming that a multilevel priority machine is used (11/04, 11/34, etc.)
2. Based on the descriptions given in Chapter 8 for the clock and A/D, write a similar program for a parallel digital I/O (e.g., DRV-11). Look up the device characteristics for the parallel I/O device in the peripherals handbook.
3. In the percolating queue algorithm given in this chapter, compute the maximum sampling speed that can be used and still avoid having a flickering display. Assume that repetitions of the display above 30 times/sec will cause the display to appear stationary.
4. For the percolating queue algorithm, draw a graph of buffer size versus maximum sampling speed for no flicker.
5. Compare the results of using the three display algorithms described in this chapter. Discuss speed, number of points used, etc. When would you want to use each algorithm?
6. Are there other efficient ways of displaying data in addition to the methods discussed?
7. What would you do to plot data slowly on an x–y plotter? Write a short routine to accomplish plotting on an x–y plotter. (Some characteristics of x–y plotters are given in Chapter 9.)
8. Figure 8.24 shows a collection of files produced by a student for performing interrupt servicing by including isr( ) and rti( ) at the beginning and end of the interrupt service routine. There is a small problem with using isr( ). Can you determine what it is?

9

Example Programs

This chapter presents several examples of programs used for specific laboratory applications. First, a program for signal averaging is presented. Second, a program for computing fast Fourier transforms and power spectra is given with an example of a method for plotting pseudo–three-dimensional displays of power spectral data. Finally, a program for producing time-interval histograms is presented. These programs represent common problems that are found in laboratory minicomputing.

9.1 SIGNAL AVERAGING

Laboratory minicomputers are often used as signal averagers. Signal averaging is an analysis methodology that extracts repetitive signals from noise. One example is the extraction of evoked potentials, recorded from the human or animal brain, from the resting electroencephalogram (EEG, brain waves). These potentials may be elicited by visual, auditory, or tactile stimuli and are normally quite small compared to the EEG. To extract evoked potentials from noise (the EEG in this case), repetitive averaging of EEG time intervals following each stimulus presentation will yield an average potential after summing many time intervals. The fundamental concept is pictorially outlined in Fig. 9.1. Signals occur in the noise after stimuli are presented, shown as pulses in this example. As successive segments are added together, the signal emerges from the noise because

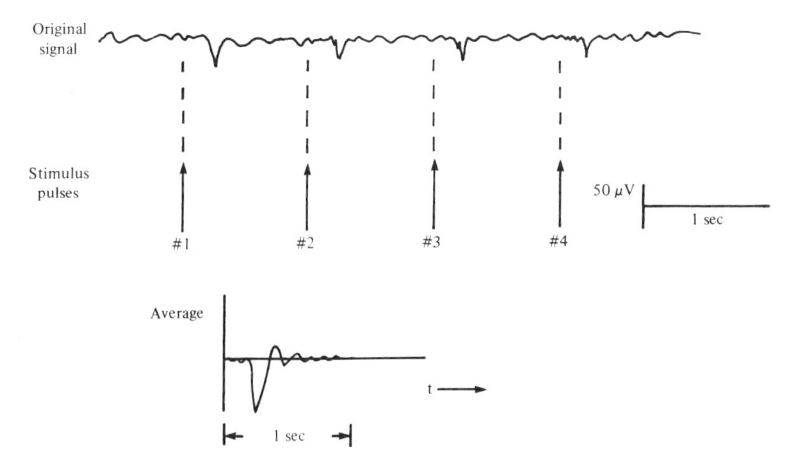

Fig. 9.1. An example of signal averaging. The original signal and stimulus pulses are shown at the top of the figure and the average waveform at the bottom. In this example the average consists of the adding segments #1, #2, #3, #4, etc.

the signal is time-locked to the stimulus while the noise is uncorrelated (read unrelated). Most signal-averaging programs do not actually average the summated signal; rather, scaling by powers of 2 is often used to control the size of the display. Since the shape of the summed waveform is the most useful information produced by averaging, actual division by the number of samples employed is infrequently carried out. If the number of samples used are a power of 2, true averaging can be easily accomplished by simply shifting the data to the right. Writing a signal-averaging program is relatively straightforward. A program outline will be given in this chapter as well as a discussion of the necessary hardware for the task.

9.1.1 Program Design Requirements

Some typical requirements for a signal averager are listed below:

1. A fast sampling rate. One thousand samples per second with a variable sampling rate option is a common requirement.
2. A variable number of sweeps. "sweeps" is the term used for each time interval following a stimulus and ending before the occurrence of the next stimulus. For example, if 100 sweeps are specified of one-sec duration, 100 one-sec segments of data are to be added together.
3. At least one channel of A/D data. Designs should include a variable number of A/D channels.

4. Display of (a) averaged and (b) sampled data as it is averaged. One should be able to select either display by typing commands on the keyboard.
5. Plotting and displaying the final average.

The ability to increase or decrease the size of the display and a method of providing a variable plot time should be included (i.e., a slow time for an *x–y* plotter; a fast time for an oscilloscope). The program should allow the user to enter the number of sweeps from the keyboard. A command facility for starting and stopping averaging is needed as well as commands for displaying, plotting, and starting over again (resetting).

9.1.2 Hardware Requirements

Figure 9.2 shows a typical system configuration for signal averaging. This example is based on the evoked-potential description discussed above. As shown, a xenon strobe light stimulus flashes and a pulse is simultaneously fed to a Schmitt trigger on the A/D, clock, or DI/O card. The signal from the scalp is amplified and sampled with the A/D. Display of the result is either on an oscilloscope or on an *x–y* plotter for final hard copy. The system can function with or without a mass-storage system.

The specific design described is based around the DEC LSI-11/2 computer with 4K words of memory and a read-only memory (ROM). Memory for program storage is necessary as well as storage for data. If only 500 points (for example) are to be sampled after each stimulus, only a

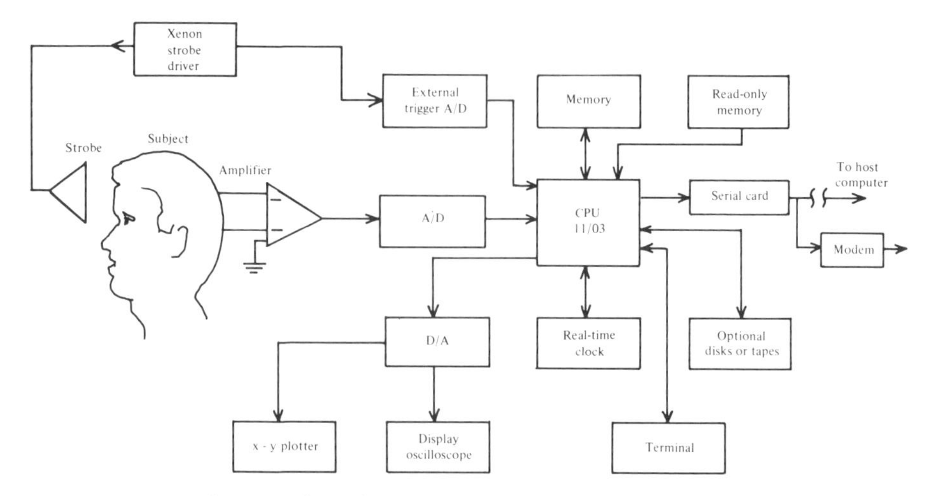

Fig. 9.2. Hardware configuration, based on an PDP-11/03, used for signal averaging.

small amount of memory is required. An A/D converter for sampling one channel (or more) is needed as are a pair of D/A converters for plotting information on either an x–y plotter or an oscilloscope. The program for averaging can be stored either in ROM or on an auxillary tape unit (e.g., the TU-58 tape cassette). A list of components for this averaging system are as follows:

(1) LSI-11/2 with 4K words of memory,
(2) ROM on a dual height card,
(3) TU-58 tape cassette (if averages are to be stored),
(4) box, power supply, backplane,
(5) A/D, D/A,
(6) serial card for terminal and TU-58, and
(7) video terminal or hard copy printer.

The system listed above will allow other programs to be run if a modest mass-storage device such as the TU-58 is purchased. If the system has access either directly or via phone line to a host system for compiling and loading programs as outlined previously, no mass storage will be required. Also, if only averaging is required, the program can be stored in ROM so that it will be always instantaneously available when the machine is turned on. If the experimental situation requires more computing than simple averaging, disks or tapes can be added. An inexpensive tape such as the TU-58 can be used for some storage but is certainly no substitute for a disk drive that can store many megabytes of data and programs. The serious laboratory experimentalist may well wish to invest in a disk drive if he has no access to a host machine. However, the extremely low cost of the TU-58 makes it an attractive storage medium for many simple laboratory computing systems.

The physical characteristics of the TU-58 drive are given in Chapter 3. The device can be connected to an LSI-11 using a common serial line interface. To read and write to the TU-58, one uses a "radial serial protocol" in which a series of characters sent in a special sequence are used to control the TU-58. For example, command packets and data packets may be sent. Each packet has a specific sequence of bytes that indicate the type of packet sent, the number of bytes in the packet, the operation code, where to read/write from/to on the tape, etc. Operation codes are simple, e.g., read, write, and position. The communication between the host computer and the TU-58 consists entirely of sending serial packets back and forth.

The entire hardware system described above is quite versatile and is a minimum configuration that can be used in a host of typical laboratory applications. The signal-averaging program is only one of many programs that could be written to run on this system. The reader is encouraged to think of other examples and to read the suggested exercises at the end of this chapter for more examples.

9.1.3 The Signal-Averaging Program

The averaging program shown in Fig. 9.3 is heavily commented. Based on information gained in previous chapters, the reader should have little difficulty in reading the code. Included at the end of the figure is a definition file "defs.h" that defines constants employed in the program.

The main body of the program is quite short, contained in only 30 lines after "main( )." This part of the code simply reads in commands and jumps to routines for servicing the keyboard commands. Other functions beside start( ), zero( ), reset( ), etc., can be easily added by simply including function names in the list. After entering the main "while" loop, a character read from the keyboard is used in the switch statement. If a value is typed in that does not have a case, the default is taken and the message "unknown command" printed. Each case contains the ASCII value for a control character. Other characters could be used or strings of characters could be matched.

A practical feature of each function is the inclusion of the "print" statement at the beginning of each function. When no code (or little code) is written in a function called by main, a dummy "stem" with only a "print" can be used to test the control flow of the program.

In "start( )" the clock is set up and started. An endless while loop keeps track of the number of sweeps and exits via 'stop( )' when the prerequisite number of sweeps are completed. After the completion of sampling, the display routine is called and the average displayed until a control e is typed.

After each presentation of the display, the data may be scaled up or down by typing the letters 'u' or 'd'. The routines up( ) and down( ) scale the data in long integer format up or down by shifting the data either left or right. With the statement

```
DA->da2 = *adpoint++;
```

the value displayed is the least-significant word of the double-precision integer. Thus to observe the average it will be necessary to scale the data down a few times if both words in the long integer are used to hold the summed average. Note that early versions of "cc" did not support long integer (32-bit integer) formats. More recent versions of "cc" are available in the program distribution packages discussed in Chapter 5.

```
#
#include "defs.h"

long adbuff[500],
*adpoint;                          /* use a double precision integer for
                                      data */

int     clkintr ();

int     count,
        numsweeps,
        sweepdone,
        alldone;

/* The various functions available in the signal averaging program are
   accessed by typing the control characters listed in the switch statement
   in main.
*/
main () {
    char    val;
    print ("signal averaging program \n\r ");
    while (1) {
        while (!(KLREAD -> tps & MASK));
        val = KLREAD -> tpb;
        switch (val) {
            case '\023':
                start ();           /* control s */
                break;
            case '\032':
                zero ();            /* control z */
                break;
            case '\001':
                reset ();           /* control a */
                break;
            case '\004':
                display ();         /* control d */
                break;
            case '\003':
                exit ();            /* control c */
                break;
            case '\020':
                plot ();            /* control p */
                break;

            default:
                print ("unknown command \r\n");
        }
    }
}

/* start() is the routine that does most of the work */

start () {

 /* print("start here \n\r"); removed after debugging */
    zero ();                       /* zero all buffers and flags */
```

Fig. 9.3. An example of a program for signal averaging written in 'C'.

```
    callC (CLINTV, &clkintr, CLPSW); /* set up clock */
    CLOCK -> clkpreset = TICKS;
                                     /* number of ticks for setting clk
                                        rate */
    adpoint = adbuff;
    while (1)
        trig_wait ();                /* wait for the input trigger */
        clockstart ();               /* then start the clock for A/D
                                        sampling */
        while (1) {
            if (sweepdone != 0) {    /* sweepdone is set in clkintr */
                sweepdone = 0;
                adpoint = adbuff;
                count = 0;
                break;               /* break gets us out of the while
                                        loop */
            }                        /* when all samples in sweep are
                                        complete */
        }
        if (numsweeps++ >= SWEEPS)   /* break when all sweeps completed */
            break;
    }
    stop ();                         /* stop the clock */
    display ();                      /* can use display command from
                                        keyboard instead */
}

clockstart () {
    CLOCK -> clkstbuf = CLOCKSTART;  /* start clock, rate determined here
                                        and by TICKS */
}


plot () {
    print ("plot here \n\r ");       /* add a wait in standard display
                                        format to create plot */
}

exit () {
    print ("exit here\n\r");         /* the user should supply his own
                                        exit depending */
                                     /* on the system being used - see
                                        chapters 7 & 8 */

}
/*      The following routines are for initialization    */

reset () {
    zero ();
}                                    /* not actually necessary, this funct
                                        is included to show construction */

zero () {
    zerobuf ();
    count = 0;
    sweepdone = 0;
    alldone = 0;
```

Fig. 9.3. (continued)

```
    numsweeps = 0;
}

zerobuf () {
    int     i;
    adpoint = adbuff;
    for (i = 0; i < 500; i++)
        *adpoint++ = 0;
}

/* stop the clock */

stop () {
    CLOCK -> clkstbuf = STOP;
}


/* wait for trigger signal on A/D and store first point away */

trig_wait () {
    AD -> adstbuf = EXTENAB;
    while (!(AD -> adstbuf & ADDONE));
    *adpoint++ = *adpoint + AD -> adin;
    count++;                              /* count the first point taken */
}


/*  the display function. When a control e is typed display() exits */

display () {
    int     i;

    while (1) {
        adpoint = adbuff;
        for (i = 0; i < 500; i++) {
            DA -> da1 = i;
            DA -> da2 = *adpoint++;
        }
        if ((KLREAD -> tpb & MASK) == 'u')
            up ();
        if ((KLREAD -> tpb & MASK) == 'd')
            down ();
        if ((KLREAD -> tpb & MASK) == '\005')
            break;
    }

}

up () {
    int     i;
    adpoint = adbuff;
    for (i = 0; i < 500; i++)
        *adpoint++ = *adpoint << 1;
}

down () {
```

Fig. 9.3. (continued)

```
    int     i;
    adpoint = adbuff;
    for (i = 0; i < 500; i++)
        *adpoint++ = *adpoint >> 1;
}


/* clock interrupt routine  */

clkintr () {
    AD -> adstbuf = ADSTART;          /* start the A/D */
    while (!(AD -> adstbuf & ADDONE));
                                      /* wait for done flag */
    *adpoint =+ AD -> adin;           /* add the current value to the
                                         buffer */
    if (count >= NUMSAMS)
        sweepdone = 1;                /* sweepdone ?,if so, set flag = 1 *
    DA -> da1 = count++;
    DA -> da2 = *adpoint++;

}


/* defs.h - definitions for the signal averaging program */


#define NUMSAMS 500                   /* the number of samples */
#define SWEEPS 10                     /* the number of sweeps */
#define CLOCK 0172540                 /* the clock address */
#define DA  0176760                   /* D/A converter */
#define AD 0176770                    /* A/D converter */
#define CLINTV 0104                   /* clock interrupt location */
#define CLPSW 0340                    /* clock psw */
#define CLOCKSTART 0111               /* value used to start clock */
#define TICKS 100                     /* 100 ticks means 500 samples in .5
                                         seconds */
#define ADSTART 01                    /* start A/D */
#define ADDONE 0200                   /* A/D done flag */
#define STOP 0                        /* used for stopping clock */
#define KLREAD 0177560                /* console terminal keyboard */
#define EXTENAB 02                    /* external enable for A/D */
#define MASK 0177                     /* byte mask */

struct {
    int     adstbuf,
            adin;
};

struct {
    int     clkstbuf,
            clkpreset;
};

struct {
    int     da1,
            da2;
```

Fig. 9.3. (continued)

```
};

struct {
    int       tps,
              tpb;
};
```

Fig. 9.3. (continued)

Obviously, this program represents the most elementary code possible. There are no sophisticated display options, nor is the ability to read in parameters included. This code is intended to provide a guide for programs the user may wish to write rather than being a complete entity in itself.

9.2 SPECTRAL ANALYSIS

There are many situations in which spectral analysis is used in laboratory situations. For example, when one requires information about the frequency of a signal, the data are often analyzed using power-spectral methods. The concept of the relationship between the time and frequency domains is easily visualized as shown in Fig. 9.4. The location of a peak in the power spectrum is directly related to the frequency of the original data and the height of the peak is proportional to the power in the original signal.

The history, algorithms, and various implementation methods for power-spectral techniques will not be discussed here since there are many excellent textbooks that fully describe theories and methods. Several of these are referenced at the end of this chapter (Bendat and Piersol, 1971; Childers and Durling, 1975; Olnes and Enochson, 1972; Rabiner and Gold,

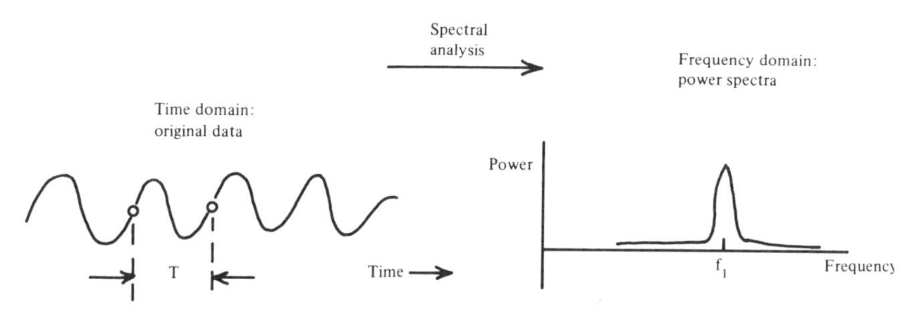

Fig. 9.4. An example of how time domain data are represented in the frequency domain. $f_1 = 1/T$.

1975). Instead, the purpose of this section is to show how one can implement a spectral-analysis program in the laboratory environment and give a specific example of one such implementation.

Common uses for power-spectral analysis are in computing the frequencies present in an electroencephalogram (EEG), speech signals, vibration data, sonar signals, etc. When dealing with data over a long period of time, it is convenient to plot the data in an easily observable and interpretable format. Section 9.3 will discuss such a plotting system.

The usual steps required to produce a power-spectral estimate as follows:

1. Sample the data using a low-pass filter to avoid aliasing.
2. Fast Fourier transform (FFT) the data, or direct Fourier transform the data.
3. Compute the power spectrum from the results of the FFT.
4. Plot the results.

Sampling can be accomplished using the methods described in previous chapters. The FFT can be computed using any number of algorithms. Useful discussions of FFT algorithms are contained in Rabiner and Gold (1975), Bendat and Piersol (1971), and Childers and Durling (1975). The power spectrum is computed from the summation of the squared real and imaginary parts of the values produced by the FFT.

The example shown next is used to compute power-spectral estimates for four channels of data. For purposes of simplification, we shall assume that the data have been sampled and are contained in a UNIX* system file with the following format:

Sample number:	Channel:	Number of data point in integer file:
Sample 0:	Channel 0	0
	Channel 1	1
	Channel 2	2
	Channel 3	3
Sample 1:	Channel 0	4
	Channel 1	5
	Channel 2	6
	Channel 3	7
	.	8
	.	.
	.	.

* UNIX is a trademark of Bell Laboratories.

The programs as presented are suitable for running on a PDP-11/34 or PDP-11/23. If the programs are to run on an 11/03, all floating-point arithmetic must be adapted to the 11/03s FPU (see Chapter 3) or simulated with all integer arithmetic. The program shown below will compute 400 one-sec power spectra from 100 sec of four-channel data in about 35 sec on a PDP-11/34 using a hardware FPU.

9.2.1 Spectral-Analysis Program

Several definitions should be supplied to the programs described:

1. NTHPOW: The total number of data points in the transform.
2. N2POW: NTHPOW = 2^{N2POW}; i.e., NTHPOW = 1<<N2POW.
3. POINTS: The number of raw data points used in the transform.

$$POINTS <= NTHPOW.$$

4. wind [POINTS]: The data window used to smooth the spectral estimate (see below).
5. corr: A correction factor used to relate sampled data to original data size. The scale or window factor of advoltbit =0.00244 is used to scale numbers produced in "dfft( )" to volt squared/Hertz values. For the 12-bit A/D example, 1/4096 = 0.00244. corr = (A/D volts/bit)^2/(nthpow)^2*window factor.
6. CHANS: The number of channels.

Two arrays for data are used: raw [POINTS] [CHANS] and data [CHANS] [POINTS]. The original data are sampled from four channels. Thus, raw [ ] [ ] contains data not separated by channel. To efficiently process the data, the floating array "data" will be defined after means are subtracted from the raw data and row and column indices reversed to allow separation by channel.

First, a cosine window is built that effectively multiplies the beginning and ending of the data by a cosine function. "corr" is a correction factor that can be included to increase the amplitude of the power estimate that is decreased by the cosine taper. Each point in the spectrum can be scaled by dividing by a window factor of 0.875 (Bendat and Piersol, 1971, p. 327). The purpose of this window is to produce an improved estimate of actual frequencies in the data. This technique provides a smoother output and reduces the variance of the output signal. It can also be accomplished after the FFT by frequency-averaging smoothing techniques. Next, means are removed; we shall assume the means have been previously computed. The data are zero padded, i.e., zeros are added to the data

```
#
/*      spectral analysis program */

#define N2POW 6                                          /* 2**6=64 */
#define NTHPOW (1<<N2POW)
#define POINTS (1<<N2POW)
#define CHANS  4
#define NUMHARM 20

main (argc, argv)
int     argc;
char  **argv;
{
    int     raw[POINTS][CHANS],
            i,
            j,
            k,
            advoltbit,
            output,
            input,
            mean[CHANS],
            ntrans,
            range;
    float   data[CHANS][NTHPOW],
            wind[POINTS],
            corr;
    char   *spectral;

    if (argc != 2)
    {
        printf ("sp fname \n");
        exit ();
    }
    if ((input = open (argv[1], 0)) <= 0)
    {
        printf ("open failed \n");
        exit ();
    }
    advoltbit =.00244;
    spectral = "testfile";
    range = (NUMHARM);
```

Fig. 9.5. Main program for use in spectral analysis. This program calls functions shown in Figs. 9.6–9.8.

until the number of points equal a power of 2. The data are then transformed and the results written to a file. For example, a program fragment would appear as shown in Fig. 9.5. This example expects the command line to contain the name of the program and the name of the data file.

```
range = range * 4;
corr = advoltbit * advoltbit / NTHPOW / NTHPOW;
if ((output = creat (spectral, 0666)) <= 0)
{
    printf ("creat failure in sp?\n");
    exit ();
}
conwin (wind, POINTS, corr);
for (i = 0; i < ntrans; i++)
{
    if (read (input, raw, (POINTS * CHANS * 2)) != (POINTS * CHANS * 2))
    {
        printf ("unexp eof in sp?");
        exit ();
    }
    for (j = 0; j < POINTS; j++)
    {
        for (k = 0; k < CHANS; k++)
            data[k][j] = raw[j][k] - mean[k];
    }
    for (j = (POINTS + 1); j < NTHPOW; j++)
    {
        for (k = 0; k < CHANS; k++)
            data[k][j] = 0.;
    }
    for (j = 0; j < CHANS; j =+ 2)
        dfft (data[j], data[j + 1], N2POW, NTHPOW, POINTS, corr, wind);
    for (j = 0; j < CHANS; j++)
    {
        if (write (output, &data[j][0], range) != range)
        {
            printf ("write failure in sp?");
            exit ();
        }
    }
}

printf ("%s\n", spectral);
if (input >= 0)
    close (input);
if (output >= 0)
    close (output);
```

Fig. 9.5. (continued)

More sophisticated arrangements would allow longer strings of file names to be entered. The function dfft( ), shown in Fig. 9.6, is used to transform two channels of data at a time. Since most data consist of real sequences (no imaginary numbers), then the symmetry properties of the Fourier

```
dfft(x,y,n2pow,nthpow,points,corr,wind)
        float *x,*y,corr,*wind;
        int n2pow,nthpow,points; {


        float xr,xi,yr,yi,*ylo,*yhi,*xlo,*xhi,half;
        register i;
        int np2;


        ylo=y;
        half=.5;
        for(i=0;i<nthpow;i++) *ylo++= -*ylo; /*conjugate y[]*/
        window(x,y,wind,points);                /*window*/
        fft(x,y,n2pow,nthpow);
        ylo=y;
        for(i=0;i<nthpow;i++) *ylo++= -*ylo;    /*conjugate again*/
        np2=nthpow>>1;
        ylo= &y[1];             xlo= &x[1];
        yhi = &y[nthpow-1];     xhi = &x[nthpow-1];
        x[0] = (x[0]*x[0])*corr;
        y[0] = (y[0]*y[0])*corr;
        for(i=1;i<np2;i++) {                    /*separate and magnitude*
                xr=(*xlo+*xhi)*half;
                yi=(*xhi-- - *xlo)*half;
                yr=(*ylo + *yhi)*half;
                xi=(*ylo - *yhi--)*half;
                *xlo++=(xr*xr+xi*xi)*corr;
                *ylo++=(yr*yr+yi*yi)*corr;
        }
        return;
}
```

Fig. 9.6. "dfft( )." This function calls a windowing function, performs the FFT and computes the power spectra from the values produced by "fft( )."

transform can be invoked to allow two sequences of numbers to be transformed using a simple transform (Rabiner and Gold, 1975, pp. 58–59).

"dfft( )" computes the spectral magnitudes of the real series stored in x[NTHPOW] and y[NTHPOW]. The algorithm employed is

1. Conjugate y[ ] since another function fft( ) computes the inverse FFT.
2. Window x[ ] and y[ ] using the window produced by conwin( ) (Fig. 9.7).
3. Transform using fft( ).
4. Conjugate y[ ].
5. Separate and compute the magnitude. (See Eq. 2.159, Rabiner and Gold, 1975.)

```
conwin(wind,points,cfact)
        float *wind,cfact;
        int points;{    /* cfact unused here, see text */
        float a,x,xa;
        double cos();
        register i,np2;

        a=.5;               /*change to .54 for hamm filter */
        xa=1.-a;
        x=6.283185307/points;
        np2=points>>1;
        for(i=0;i<points;i++) *wind++=a+xa*cos(((i-np2)+.5)*x);
        return;
}
```

Fig. 9.7. "conwin( )." The cosine windowing function.

The spectral values are returned in x[ ] and y[ ] and in Fig. 9.5 are written to an output file.

The functions called by dfft( ) are window( ) and fft( ). Window( ) simply multiplies the function computed by conwin( ) times the data and fft( ) computes the inverse fft. The algorithm shown is known as the Sande–Tukey method (Childers and Durling, 1975) and uses computed sine functions. Additional speed can be achieved using a sine look-up table. A bit reversal routine is included. The reader who is interested in the details of how the algorithms for these programs operate should examine the references given at the end of the chapter.

9.2.2 Implementation Pitfalls

A variety of common problems face the user of spectral-analysis programs including aliasing, leakage, and the picket-fence effect. These three problems are discussed in some detail by Childers and Durling (1975). Aliasing can be avoided by sampling at greater than twice the highest frequency in the signal. Leakage is minimized by using a smoothing window on the data [conwin( ) in programs described above]. Leakage refers to a phenomenon produced by analyzing finite data segments. Truncation of data at the beginning and end of a data epoch produces side lobes in the power spectrum termed leakage. Use of a window to taper the beginning and ending of an epoch reduces the error. The picket-fence effect is concerned with the problem of representing the spectral density at each frequency where each representation is not uniform, thus creating a line of responses along the frequency axis that look like a picket fence. The solution is to move the slats in the fence closer together by zero padding

```
window(real,imag,wind,points)
        float *real,*imag,*wind;
        int points;{

        register i;
        for(i=0;i<points;i++){
                *real++=* *wind;
                *imag++=* *wind++;
        }
        return;
}

bitrev(x,y,nthpow,n2pow)
        float *x,*y;
        int nthpow,n2pow; {

        int i,k,a1,a2;
        char tag[256];
        float i1,r1;

        for(i = 0; i < nthpow; i++) tag[i] = 0;
        for(a1 = 1; a1 < nthpow; a1++){
                if((!tag[a1]) && ((a2 = reverse(a1,n2pow)) != a1)) {
                        i1 = y[a1]     ; r1 = x[a1];
                        y[a1] = y[a2] ; x[a1] = x[a2];
                        y[a2] = i1     ; x[a2] = r1;
                        tag[a2] ++;
                }
        }
        return;
}

reverse(a1,n2pow)
        int a1,n2pow;{

        register i,j,k;
        i = a1;
        j = 0;
        for(k = 1; k <= n2pow; k++){
                j = (j<<1)|(i&1);
                i = (i>>1);
        }
        return(j);
}
```

Fig. 9.8. A collection of functions used for computing the FFT.

the data up to the next power of 2. This ability is provided in the programs listed by manipulating N2POW and POINTS.

To use the algorithms exactly as specified above, either a PDP-11/23, PDP-11/34, or other PDP-11s with an FP-11 hardware compatible

```
t(x,y,n2pow,nthpow)
      float *x,*y;
      int n2pow,nthpow;{
      float sinsca,cosca,c1,s1,sine2,c2,s2,c3,s3,r1,r2,r3,r4,i1,i2,i3,i4;
      int pass,seqloc,n4pow,nxtlth,length,j1,j2,j3,j4,j;
      double scale;
      double sin(),cos();
      n4pow = n2pow>>1;                    /* # radix-4 passes needed */

perform radix-4 passes, if any

      for(pass = 1; pass <= n4pow; pass++) {
              nxtlth= 1<<(n2pow-(pass<<1));
              length=nxtlth<<2;

 the next 8 statements could be replaced by a sine lookup table
(except for the 'for' statement)

              scale=6.283185307/length;
              sinsca=sin(scale);
              s1= -sinsca;
              c1=cosca=cos(scale);
              for(j=0;j<nxtlth;j++){
                      sine2=s1*cosca+c1*sinsca;
                      c1=c1*cosca-s1*sinsca;
                      s1=sine2;
                      c2=c1*c1-s1*s1;
                      s2=c1*s1+c1*s1;
                      c3=c1*c2-s1*s2;
                      s3=c2*s1+s2*c1;
                      for(seqloc=length;seqloc<=nthpow;seqloc=+length){
                              j1=seqloc-length+j;
                              j2=j1+nxtlth;
                              j3=j2+nxtlth;
                              j4=j3+nxtlth;
                              r1=x[j1]+x[j3];
                              r2=x[j1]-x[j3];
                              r3=x[j2]+x[j4];
                              r4=x[j2]-x[j4];
                              i1=y[j1]+y[j3];
```

Fig. 9.8. (continued)

floating-point unit must be used. To run the programs on an 11/03, two or more routes are possible. One is to compile the program on a MINIUNIX* system using the cc compiler configured for the PDP-11/40, which has a floating-point unit compatible with the 11/03. Alternatively, floating-point arithmetic may be simulated on the 11/03 and functions may

* MINIUNIX is a trademark of Bell Laboratories.

```
                                        i2=y[j1]-y[j3];
                                        i3=y[j2]+y[j4];
                                        i4=y[j2]-y[j4];
                                        x[j1]=r1+r3;
                                        y[j1]=i1+i3;
                                        if(j!=0){
                                                x[j3]=c1*(r2-i4)-s1*(i2+r4);
                                                y[j3]=s1*(r2-i4)+c1*(i2+r4);
                                                x[j2]=c2*(r1-r3)-s2*(i1-i3);
                                                y[j2]=s2*(r1-r3)+c2*(i1-i3);
                                                x[j4]=c3*(r2+i4)-s3*(i2-r4);
                                                y[j4]=s3*(r2+i4)+c3*(i2-r4);
                                        } else {
                                                x[j3]=r2-i4;
                                                y[j3]=i2+r4;
                                                x[j2]=r1-r3;
                                                y[j2]=i1-i3;
                                                x[j4]=r2+i4;
                                                y[j4]=i2-r4;
                                        }
                                }
                        }
                }
/*
*  radix-2 pass if any
*/
        if(n2pow&1) for(j=0;j<nthpow;j=j+2) {
                r1=x[j]+x[j+1];
                x[j+1]=x[j]-x[j+1];
                x[j]=r1;
                i1=y[j]+y[j+1];
                y[j+1]=y[j]-y[j+1];
                y[j]=i1;
        }
/*
* a bit reversal and all's done
*/
        bitrev(x,y,nthpow,n2pow);
        return;
}
```

Fig. 9.8. (continued)

be written to multiply x and y and return the result in z; i.e., $z = \text{mul}(x,y)$. Of course, floating-point emulation will produce programs that require many seconds to compute the set of spectra required for the CSA plot. Perhaps the most reasonable method for using the spectral analysis program is to sample data on an 11/03 and transfer the raw data to a larger PDP-11 that can use a hardware FPU to speed up the computations.

9.3 COMPRESSED SPECTRAL ARRAY PLOTTING

A convenient representation for plotting the power spectra produced in programs similar to the program shown in the previous section is the so-called compressed spectral array (CSA). The CSA is essentially an isometric pseudo–three-dimensional plot for displaying power-spectral data. Other types of data can be also easily displayed using the plotting algorithm. The program shown in Fig. 9.9 plots each spectral line behind the previous line and suppresses any data that fall behind previously plotted data. The first line is plotted, e.g., as in Fig. 9.10. Next, the second line is plotted with an increment (INC) added to each data point. When values on the second line plus the increment are less than the value on the first line, points on the first line are plotted, e.g., as in Fig. 9.11. Successive lines are plotted similarly until all lines specified in NUMLINES are plotted, e.g., as in Fig. 9.12. The x and y output of the program is via the D/A, suitable for running on the 11/03. The CSA can be plotted easily on an oscilloscope, but must be modified for use with an x–y plotter. For example, at the end of each line a wait should be inserted to allow the pen to return physically to the beginning of the next line. Further, the pen must be lifted from the paper before the new x–y position is given. This type of operation can be accomplished by use of a digital I/O. For example, the DRV-11 parallel I/O can be used for digital output. The registers may be defined as

```
struct {int control_status,
            output,
            input;}
```

and the DI/O address as

```
#define DI/O 0167760
```

To produce a voltage on bit zero of the output register, a statement could be added to set the output:

```
DI/O->output = 01;
```

or a zero to clear the output:

```
DI/O->output = 0;
```

The change in voltage levels produced on this output bit could be used to drive a relay that physically raises or lowers the pen.

The program, as shown, continuously adds the value INC to each line. Obviously there are physical limitations to continuing to add values in this

```
#
#define INC               10          /* vertical increment */
#define NUMPOINTS         80          /* examples */
#define NUMLINES          20
#define DA                0176760

struct
{
    int     da1,
            da2;
};
            int x[NUMLINES][NUMPOINTS];

csaplot ()
{

    int     i,
            j,
            l,
           *pp,                       /* pp = pltbuf pointer */
           *dp,                       /* dp = data pointer */
            inc,
            pltbuf[NUMPOINTS];

    inc = INC;                        /* vertical increment
                                      */
    dp = x;                           /* data pointer = X array address
                                      */
    pp = pltbuf;
    for (l = 0; l < NUMPOINTS; l++)
    {
        *pp++ = *dp;                  /* plots 1st  */
        DA -> da1 = l;                /* line       */
        DA -> da2 = *dp++;
    }
    for (i = 0; i < NUMLINES: i++)
    {
        pp = pltbuf;
        for (j = 0; j < NUMPOINTS: j++)
        {
            if ((*dp + inc) >= *pp)
                *pp = *dp++ + inc;
            else
                dp++;
            DA -> da1 = j;
            DA -> da2 = *pp++;
        }
        inc =+ INC;
    }
}
```

Fig. 9.9. An outline of a program for plotting compressed spectral arrays.

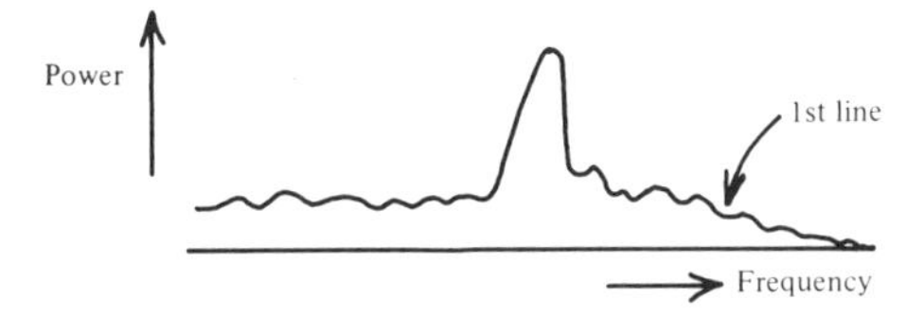

Fig. 9.10. A plot of the first line of the CSA.

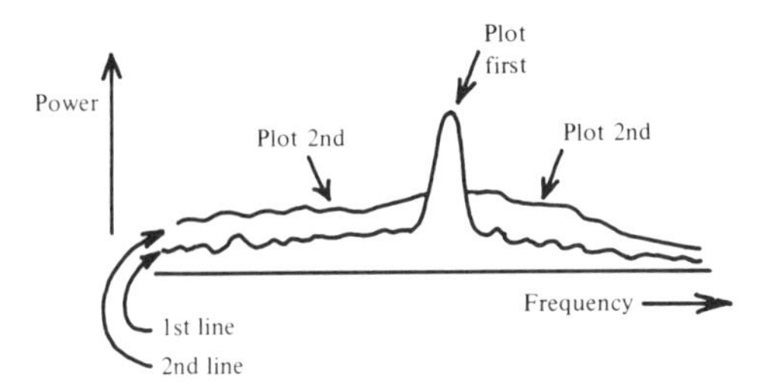

Fig. 9.11. How the second line in a CSA is plotted with respect to the first.

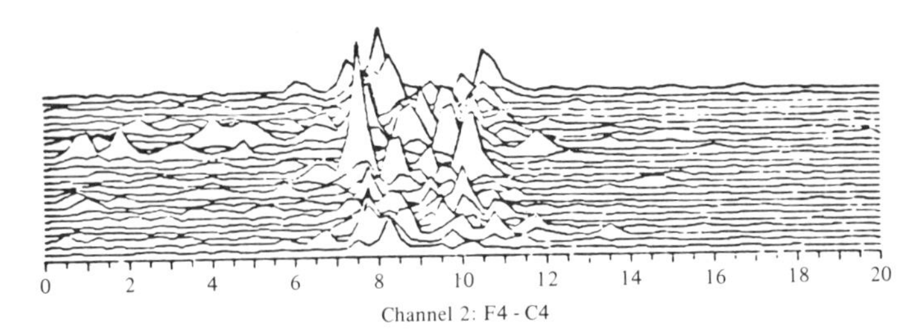

Fig. 9.12. A complete compressed spectral array plot.

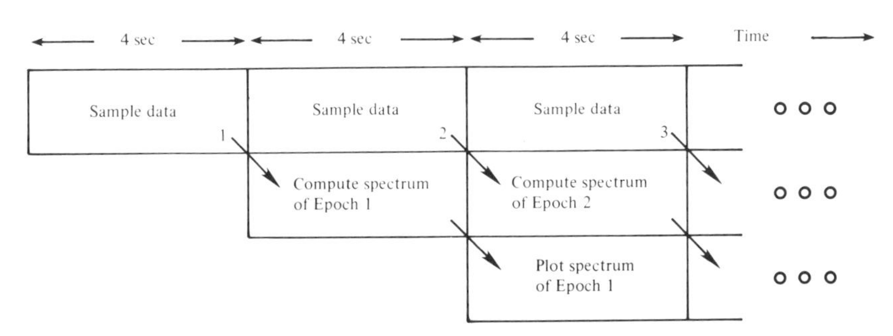

Fig. 9.13. An example of how sampling data, computing spectra, and plotting can be contemporaneously accomplished. This example shows 12 sec of time. Arrows indicate the flow of data from one process to another. Data sampled in 1 is analyzed and plotted in the other 4-sec segments labeled with a 1.

manner. Sooner or later the 12-bit D/A will wrap around and produce a negative value (i.e., 3777 = positive number; +10 = 4007 = a negative number). Solutions are (1) to contain the y-axis values in a set range, or (2) to let the y axis "wrap around" and begin plotting at the bottom of the figure after it completes plotting at the top.

9.3.1 Integration of Sampling, Power-Spectral Computation, and Plotting

For real-time laboratory applications, it is convenient to integrate sampling, the power-spectral computations, and plotting into one package. The most straightforward implementation method is to perform sequentially these three functions. However, this method is not as time-efficient as techniques in which the three operations contemporaneously proceed. Consider Fig. 9.13, which diagrammatically shows how such operations can proceed. As an example, using a 4-sec transform, data are sampled for 4 sec. During the next 4 sec, the spectrum of the first 4 sec is computed. Then, during the subsequent 4 sec, the spectrum from the previous 4 sec is plotted in CSA format, while the previous data-epoch sampled is spectrally analyzed. Thus three operations can be conducted effectively at the same time. In this example, a plot of the spectral data begins being plotted only 8 sec after initiation of sampling. However, this type of real-time computing cannot be carried out unless the FFT is sufficiently fast to be completed in less than 4 sec so that time remains for sampling and plotting.

Trying to implement the above strategy on a PDP-11/03 using a floating-point algorithm is futile due to the slowness of the floating-point simulation. An alternative is to use integer arithmetic throughout for speed. If accuracy is a problem, long integer format is an alternative. PDP-11s with floating-point hardware will easily allow implementation of such strategies.

9.4 TIME-INTERVAL HISTOGRAM

Figure 9.14 displays an example of code that can be used for producing a rudimentary time-interval histogram. This code is only partially complete and can be supplemented and modified using routines and ideas previously presented. Consider the random stream of pulses shown in Fig. 9.15. In various research and applied problems, it is often of interest to compute the time-interval histogram of such data. For example, when

```
include "defs"                  /* definitions for clock, terminal, A/D
                                   and D/A */
define BINWIDTH 100             /* number of counts in a bin, for
                                   example at 10000 Hz, 100 counts
                                   produces .01 second bin */
define NUMBINS 100              /* the number of bins in the histogram
                                */
define LEFTSCOPE -2000          /*  leftmost point to be plotted on
                                   oscilloscope */
define RIGHTSCOPE 2000          /* rightmost point to be plotted on
                                   oscilloscope */
define STARTAD 02               /* enable external events to be sensed
                                   with A/D */
define STARTCLK 023             /* enable clock at 10 khz and up
                                   counting */
nt      adserv ();
har    *pt1,
       *pt2;                    /* clock counter value holders */

nt      bin[1000],
        firsttime;

ain ()

   setup ();                    /* setup clock and wait for control s
                                   to begin */
   while (1)
   {
       if (KLREAD -> tkb & MASK == 03)
       {
           CLOCK -> status = 0;
                                /* turn off clock before exiting */
           exitprog ();
       }                        /* 3 = control c */
       display ();
   }


xitprog ()

   rtexit ();                   /* use if running 11/03 under RT-11 */
*      exit();    use if running MINIUNIX */
*      halt();  use if running standalone or do not use */
*      see figure 8.24 for functions halt() and rtexit() above */


etup ()

   pt1 = pt2 = firsttime = 0;   /* initialize counter values holders */
   KLPRINT -> status =& ~0100;  /* turn off printer interrupt */
   if (callC (ADINTV, &adserv, ADPSW) < 0)
   {                            /* use A/D for sensing input pulses */
       print ("error in setting up A/D \r\n ");
                                /* use 11/03 print routine - see
                                   chapter 6 */
```

Fig. 9.14. An example 'C' program for computing and plotting a histogram.

```
        exitprog ();
    }

    print (" Hit control s to begin \r\n");
    while (KLREAD -> tkb & MASK != 023);
                                        /* wait for control s */

    CLOCK -> status = STARTCLK;         /* start the clock ticking and let it
                                           run */
}

adserv ()
{
    int     time,
            delta_t;                    /* t1 is first time value, t2 is secon
                                           time value  used */
    time = CLOCK -> counter;            /* delta_t is used to get the clock
                                           counter value */
    if (firsttime != 0)
    {
        pt2 = time;
        if (pt2 < pt1)
            delta_t = 0177777 - pt1 + pt2 + 1;
                                        /* 0177777 is used as the largest valu
                                           the counter obtains before wrapping
                                           around */
        else
            delta_t = pt2 - pt1;
                                        /* note that pointers to characters ar
                                           used so that unsigned arithmetic */

        bin[delta_t / BINWIDTH]++;
                                        /* can be used. The counter produces
                                           unsigned numbers while normal
                                           arithmetic is signed  */
        pt1 = pt2;                      /* pt1 is replaced for next time
                                           through */
    }
    else
    {
        pt1 = time;                     /* first time through only */
        firsttime++;
    }
}

display ()
{
    int     i,
            j;
    j = 0;
    for (i = LEFTSCOPE; i <= RIGHTSCOPE; i =+ (4000 / NUMBINS))
    {
        DA -> da1 = i;                  /* x axis */
        DA -> da2 = bin[j++];           /* y axis */
    }
}
```

Fig. 9.14. (continued)

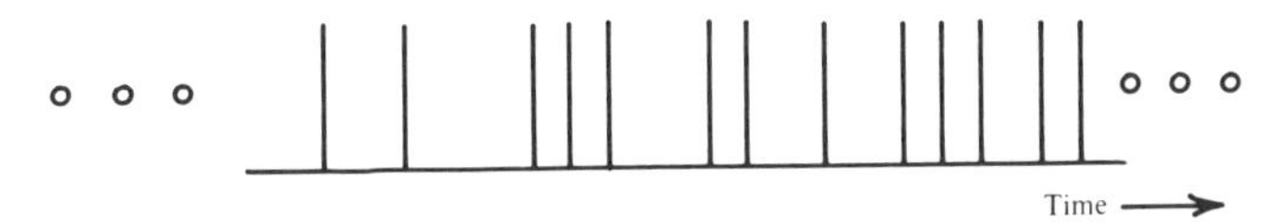

Fig. 9.15. A randomly spaced sequence of pulses.

psychologists or neurophysiologists measure the firing rates of nerves, they are often interested in producing a histogram such as the example shown in Fig. 9.16, which summarizes the pulse activity being studied. In Fig. 9.16, the data shown would indicate that there were many occurrences of pulses with time intervals between the pulses of 20–40 msec while there were relatively few with time intervals of 70 msec or greater. Thus the method can give an easily interpretable output, providing average frequency and distribution information. Other examples in which histograms of this type are used are reaction time measurements in which subjects press keys after some stimulus. A variant on the time-interval histogram theme is the poststimulus histogram (PSH) in which the times after a specific stimulus are measured rather than the time between data pulses.

The method employed in this example for extracting a time-interval histogram from a random stream of pulses uses the clock and the external trigger on the A/D. The input data stream from the external source should be buffered using a Schmitt trigger to allow a threshold to be set below which pulses will be ignored. This procedure avoids the problem of having random noise treated as data.

The programming concept is straightforward. The clock is set to tick at a predetermined rate (e.g., 10 kHz) and after typing a "control s," counting begins. The A/D external input is set to allow sensing of the pulse inputs. Alternatively, one input on a digital parallel I/O could be used to sense the pulses. One programming approach would be to start the clock after each pulse and stop the clock when the next pulse occurs. However,

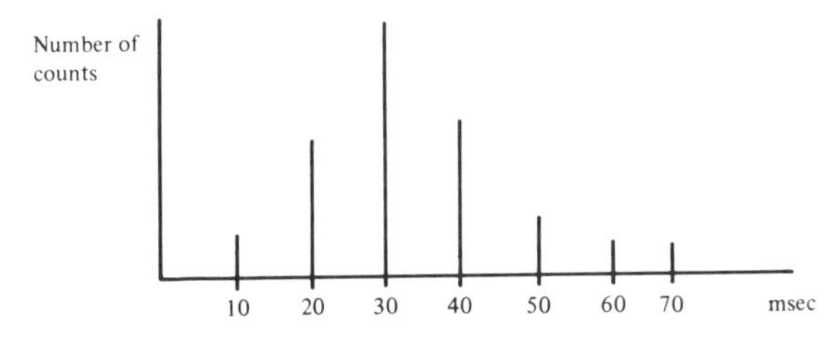

Fig. 9.16. An example of how a histogram of a sequence of randomly spaced pulses might appear.

since the clock can be read while counting, it is easier to simply let the clock run freely and read the clock counter on each interrupt of the A/D external trigger. Using the standard set-up mechanisms and definitions in Chapter 8, the reader should be able to tailor this code to his own desired usage. In the interrupt routine, the clock is read on each external interrupt (i.e., whenever a pulse from the external data stream causes an interrupt). The first time an interrupt occurs, the count at which it occurs is stored in variable pt1. On successive interrupts, pt2 is filled with the count and delta_t is computed as the difference between pt2 and pt1. "pt1" and "pt2" are defined as pointers to characters so that arithmetic on these numbers will be unsigned. This step was taken since the clock counter produces unsigned numbers. Arithmetic on pt1 and pt2 is straightforward since pointers are stored as integers. Of course, if the count (pt2) exceeds 65,535(2^{16}), pt2 − pt1 will be negative. Thus delta_t must be computed differently when wraparound occurs (i.e., the clock resets to zero after 65,535). Since unsigned arithmetic is employed, the difference between pt1 and 65,635+1 can simply be added to the pt2 value to compute the correct number of counts.

After each delta_t is computed, the "bin" corresponding to that delta_t is incremented. In histograms, the counters that indicate the number of occurrences in a given interval are called "bins." If BINWIDTH is 100 counts and delta_t is less than 100, bin [0] will be incremented. Similarly, if delta_t is between 100 and 200, bin [1] will be incremented, and so forth. Note that there is no way to determine if the clock has overflowed. At 10,000 Hz, the clock counter will reset approximately every 6.5 sec (65536/10000). If longer times between pulses are required, the only alternatives available with this clock are to feed the clock a slower periodic pulse on its external input or use the 60-Hz line frequency. Other real-time clock models often have selectable count rates that are considerably slower than the KW11-P (e.g., 1 kHz, 100 Hz).

The display routine simply sweeps the x axis of the scope while displaying the number of counts in the bins on the y axis. As shown, the display will produce a point for each bin on the scope corresponding to the number of counts in the bin. If a vertical bar is desired, the user must supply this code. The display is continuously executed in main after first checking to see if a control c (to indicate the user's desire to exit) has been typed on the console keyboard. "exit( )" is a routine that can take several forms as shown depending on the software configuration. Three methods for exiting are shown, including simply halting the computer.

A facility to stop data taking and display the results should be added to the program as well as scaling routines, such as the up( ) and down( ) functions presented earlier in this chapter. The computation of which bin to increment can be conducted outside of the interrupt routine by simply storing each clock value in a list that is accessed when the interrupt routine is not active. Use of this scheme will permit pulses with shorter interpulse times to be detected. As with most laboratory programs, the user will find it convenient to add other facilities as they are required. A modular approach for selecting between options, as presented for the signal averaging program, is recommended.

EXERCISES

Given below are several exercises, typical of problems found in the laboratory.

1. Write a program to compute the frequency histogram of sampled data using a zero-crossing algorithm. For example, consider the variable frequency waveform shown in Fig. 9.17. The frequency histogram of this signal could be predicted to look somewhat like that in Fig. 9.18.
   (a) Compute the time between successive zero crossings of the waveform and calculate the reciprocal ($f = 1/T$).

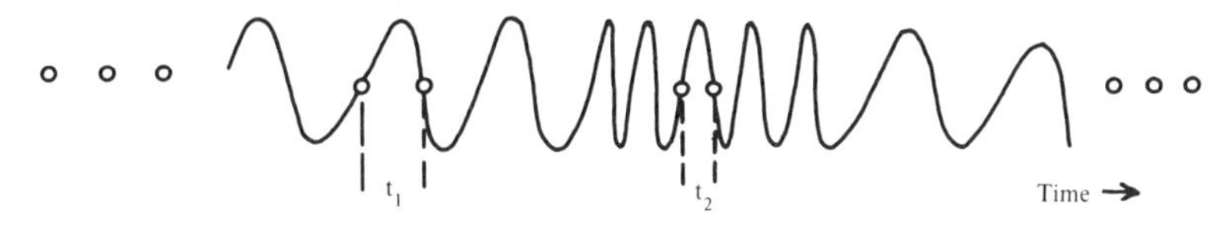

Fig. 9.17. A sinusoidal waveform. t_1 and t_2 are two time intervals between zero crossings in the data. For the purposes of comparison with Fig. 9.18, consider that the data continue with these same characteristics for a longer period of time than shown here.

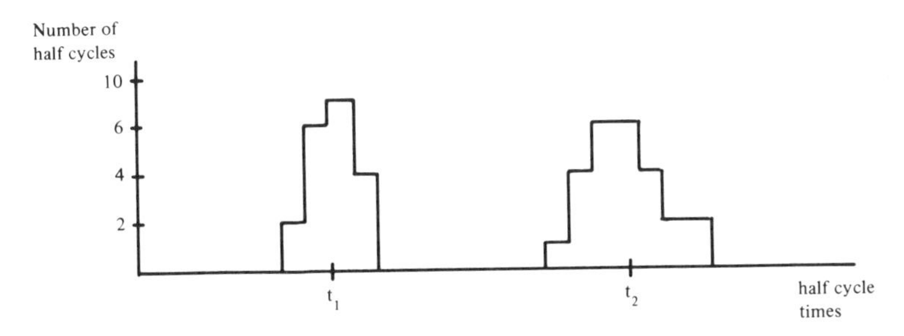

Fig. 9.18. A representation of a histogram of zero-crossing times of a wavetrain such as the one shown in Fig. 9.17. The number of half cycles in the sinusoids are counted and cluster around t_1 and t_2.

(b) Plot the histogram of frequencies on the oscilloscope and on an x–y plotter.
(c) Compare the results you get with a spectral plot of the same data.
(d) Is this problem any different from the example given in Section 9.4?

2. Rapidly sample a signal from a microphone and play the signal back via a D/A converter to a loudspeaker.

   (a) What is the sampling rate necessary to produce understandable speech? Compute the rate required for both (1) low-pitched and (2) high-pitched voices.
   (b) Using a digital filter algorithm, filter the data prior to feeding the output signal to the speaker. What is the effect of high, low, and bandpass filtering?

3. A program is required to control an experiment in which a stimulus is presented to a subject and the subject's response is to be recorded. A typical experimental paradigm in psychological laboratories is to flash a light (or present a number, e.g.) and ask the subject to choose between two or more alternatives. Flashing the light can be controlled by a DI/O and a relay. Recording the subject's response can be monitored by the DI/O as well. Write a program for the following test situation.

   (a) Present a string of numbers on the oscilloscope briefly (e.g., $\frac{1}{4}$ sec).
   (b) Follow the presentation of the above string with one number that *is* or *is not* contained in that string. The subject is to indicate whether the number was in the string by pressing one of two buttons (yes, it was; no, it was not). The time for presentation of the test number after the first string should be variable.
   (c) Count the number of correct and incorrect responses the subject makes.
   (d) Compute the subject's average reaction time (the time it takes him to press the button). You may wish to consider how to compute the standard deviation of the responses, as well.
   (e) Make all presentations random or pseudorandom.
   (f) Present a table summarizing the subject's responses.

4. Biofeedback is an intriguing area. Consider the control of heart rate. Write a program that samples the EKG signal from the heart, computes the heart rate, and feeds the value continuously back via a D/A converter to a voltage-controlled oscillator and an amplifier/speaker. As a subject listens to the tone produced, he should be able to raise

or lower his heart rate. Similar experiments can be conducted with blood pressure, ear drum temperature, gastric motility, EEG signals, etc.

5. Many game programs can be written for minicomputer systems. Using the D/As and an oscilloscope, write a lunar lander game in which a lunar module is guided toward a destination by moving a joy stick. You will need to construct a joy stick from two potentiometers and batteries and feed the derived voltage to two channels of the A/D. The joy stick is used to control the x–y motion of the lunar lander. Specify the constraints of your problems, i.e., are there fuel limitations, rate settings, etc.?
6. Write a program to simulate driving. A simple version would be to present a box on a screen that moves back and forth. The subject's task is to move a steering wheel to keep an "x" within the box.

   (a) After implementing this system, add a feature that allows computation of the percentage time that the "x" is in the box.
   (b) Make the task adaptive, i.e., make the box move faster until the subject cannot keep up with it.

7. An automated digital circuit tester can be constructed by controlling a digitally controlled power supply with the DI/O. A simple operational amplifier tester or transistor tester can also be constructed for producing characteristic frequency curves, etc. How can this be done?
8. Design a program to track a satellite with a moving antenna (for example, the amateur radio satellite). In brief, you will need to control antenna errors, determine if the antenna is locked in on the satellite by checking signal strength, and optimize antenna movement by using a feedback system.
9. Improve the signal-averaging program presented in Section 9.1 including enhanced display and control options.
10. Implement the modifications suggested in the text to the CSA plot program to allow plotting with an x–y plotter. Assume you have a standard x–y plotter with adjustable x and y gain and a digital raise/lower input for the pen.
11. Consider reworking the above problem for a digital incremental point plotter. The characteristics of such a plotter are that there are typically six digital inputs:

    (1) raise the pen; (2) lower the pen; (3) move the pen 0.005 (or 0.01 or other increment) in. in the x direction; (4) move the pen 0.005 in. in the $-x$ direction; (5) move the paper 0.005 in. $+y$; (6) move the paper 0.005 in. $-y$.

What steps would you take to use such a plotter? What would be the drawbacks and advantages compared to a standard analog x–y recorder?

12. As a project, combine a sampling program with the spectral-analysis programs given in Section 9.3. First, demonstrate that a sine wave input from a signal generator produces a peak in the output spectrum at the appropriate frequency. Next, implement the CSA plotting algorithm and plot spectra for 1- and 4-sec segmented data. What differences do you observe for plots using 1- and 4-sec data?

REFERENCES

Bendat, J. S., and Piersol, A. G. (1971). "Random Data: Analysis and Measurement Procedures." Wiley (Interscience), New York.

Childers, D., and Durling, A. (1975). "Digital Filtering and Signal Processing." West Publishing Company, St. Paul, Minnesota.

Olnes, R. K., and Enochson, L. (1972). "Digital Time Series Analysis." Wiley (Interscience), New York.

Rabiner, L. R., and Gold, B. (1975). "Theory and Application of Digital Signal Processing." Prentice Hall, New Jersey.

Appendix A
Example and Discussion of Methods for Bringing up the UNIX* Operating System on PDP-11s

This section supplements the information in the documentation supplied with the UNIX system on bringing up a UNIX system on a PDP-11. No code from the operating system is reproduced in order to conform with license restrictions. Comments are made on the various sets of documentation supplied to UNIX system licensees. An example of how to assemble and compile source code is given.

A.1 COMMENTS ON "SETTING UP UNIX"

The first document in the set of documents supplied with UNIX time-sharing system is a paper entitled "Setting up UNIX." This document provides a concise description of how to set up a UNIX system starting with an image of the UNIX system on magtape and one of several disk drives. Unfortunately, the disk drive the user has may not be one of those for which there are bootstraps already available. Thus, one must consider other mechanisms for bringing a UNIX system into being. This set of comments is intended to be read after the potential UNIX system user has completely read the UNIX system documentation.

* UNIX is a trademark of Bell Laboratories.

Situation 1. You have one of the disk drives and magtape units specified in the UNIX setup document. If you follow the instructions given, the UNIX system will come up in a minimal configuration. Once a running system is obtained, the specific devices you want on the system can be added.

Situation 2. You have a disk drive on a system that is not one of those specified in the original manual. For example, boots are supplied for tape to RK05, RP03, and RP04 disks. However, suppose you have only an RL01 or RK07. All you can do is to find a PDP-11 system somewhere that has either the RK05, RP03, or RP04, borrow the machine for a while, and construct your system on tape before bringing it back to your PDP-11. Another alternative is to borrow one of the abovementioned drives, plug it into your system, bring up the UNIX system and build a new UNIX system that will run on your system.

From this point forward we shall assume that you have either borrowed a computer system or obtained a disk drive that will fit on your computer. The tape from the Bell System contains a brief bootstrap at the beginning that can be accessed using the code indicated in the "Setting up UNIX" document. For the TU-10 magtape drive, these are the six instructions beginning with "12700." These instructions can be placed in memory and executed to boot the tape. Other devices have a similar set of instructions. The reader who wants to know what the instructions do will find that information in boot (VIII). The boot itself, called mboot, is found in /usr/mdec, and resides in the second block of the tape. The purpose of entering six (or so) instructions into memory is to read this block into memory starting at location zero. When another tape drive is used, the user must determine what instructions are to be entered into memory for the boot. You can use the TU-10 boot as a model. When the block is read into memory and restarted at zero, an "=" is printed on the console. Then one of the programs named tmrk, tmrp, htrk, etc., can be typed. 'tm' refers to a TU-10, 'ht' to a TU-16, 'rk' to an rk05, etc. These programs follow the boot and simply copy a binary representation from tape to disk. Once the copy is made, the disk itself can be booted (for example, using the ROM bootstrap).

A.1.1 Modifying the Boot

You can modify the boot programs so that you can boot your system from tape. For example, if you have an RK06, it would be convenient to be able to type tmr6 and transfer a tape to a disk. This notion is also very useful in backing up a system. Once tmr6 is made for your RK06-based system (for example), you can tp the file to tape, i.e., % tp rm0cv tmr6. 'tp' will write the initial boot onto tape. Other boots could be put on tape, as well, e.g., tmrl (for an rl01), tmr2 (for an rl02), etc. When you are running the UNIX system on disk and wish to back it up, the "dd" command can make an image copy. For example, to copy 410 blocks from disk to tape

% dd if=/dev/rhpl of=/dev/rmt0 bs=33792 count=410

This command copies an entire rk06 to tape—see dd(I). Once copied, it can be retrieved using the tmr6 program produced above. Next,

(1) toggle in the boot,
(2) tape moves—reads in mboot,
(3) type in tmr6,
(4) mount the dd copy of the UNIX system,
(5) run tmr6: tmr6 asks for (in blocks)

(a) disk offset
(b) tape offset
(c) count.

Supply 0, 0, 4000 to the three questions above and the tape will be copied to the disk. You can then halt the machine and boot the disk. Thus, with the two tapes:

(1) one created with 'tp', containing a boot and
(2) one created by 'dd', one can easily back up the system in magtape bootable form.

A.1.2 Other Means of Backup

If you have several disks on a system, you can make any of them bootable. For example, if you have an RK05, you can use the Bell-supplied tape to bring up a UNIX system in the original configuration. Then you can use 'dd' to copy a backup tape to the disk that may have failed or needs a new system on it. Once copied the system can be booted. For example,

```
Step 1. control boot-> $RK ret [RK05]
     2.                @unix
     3.                % [dd to RP04 here]
     4.                % [sync]
     5. control halt
     6. control boot-> $DB [RP04]
     7.                @unix
```

A.2 COMMENTS ON BUILDING A SYSTEM

The UNIX system documentation that explains the methods used for bringing up a system is excellent, read it all. To clarify further, an example of bringing up a system is given below:

(a) To proceed you must have a running system as outlined in the previous section.

(b) System operation.

(1) Compile and create libraries of system and handlers. The entire system source is continued in /usr/sys. In /usr/sys/ken:

```
cc  -c -o *.c
ar  r . ./lib1
rm  *.o
```

Now the system is in /usr/sys/lib1. In /usr/sys/dmr:

```
cc  -c -o *.c
ar  r . ./lib2
rm  *.o
```

puts all the handlers in lib2 that are referenced in "c.c." and "l.s". Before compiling the handlers, it may be necessary to modify some of the parameters contained in the original distribution. For example, one can edit /usr/sys/dmr/kl.c to indicate the number and kind of serial devices on your system. When doing this check the addresses of the devices as well (see a peripherals or terminal handbook). If only the kl.c modification is made after the generation of lib2, recompile kl.c and replace the old version of lib2 with the new one.

```
cc -c -o kl.c
ar rv . ./lib2 kl.o
```

Follow the same procedure for each modification that is necessary.

(2) If you are bringing up a new system, the mkconf configuration program can be used. In usr/sys/conf, the following steps can be taken to create a configuration program for either PDP-11/40-type machines or PDP-11/45-type machines.

```
as m40.s          machine language support
mv a.out m40.o
: as m45.5
```

```
: mv a.out m45.o
: cc sysfix.c          for 11/45, 70
: mv a.out sysfix
cc mkconf.c
mv a.out mkconf
```

Commands preceded by ":" are for 11/45-type systems and are interpreted by the shell as comments. Included in this category are PDP-11/45s, 55s, 70s, and 44s. In the 11/40 category are 11/34s, 11/40s, 11/23s, 11/60s. To run the configuration program, simply type

```
mkconf
rk
tm
tc
done
```

to create a system with an rk05, magtape (tm), and dectape (tc). This program generates c.c and l.s, files with names and addresses of handlers (c.c), and interrupt vector locations (l.s). These files can be conveniently edited when further modifications to the system are made. When adding device handlers, be sure the size of the system is not increased beyond 24K. The size can be controlled by resetting the number of "in-core" buffers, set in "/usr/sys/param." For example, when we added a versatec printer plotter to our system, the number of buffers was reduced from 18 to 17. The first six PAR's in the memory management unit each point to 4K words ($6 \times 4 = 24$). If more partitions are requested, the next PAR will be overwritten, causing all kinds of havoc.

(3) Compile, link programs, and copy to the system.

```
           cc -c c.c
           as l.s
           ld -x a.out m40.o c.o . ./lib1 . ./lib2
           :as data.s low.s
11/45 or   :ld -x -r -d a.out m45.o c.o . ./lib 1 . ./lib2
   70      :nm -ug
           :sysfix a.out x
           :mv x a.out
           size a.out /rkunix
           mv a.out /rkunix
```

Now the system is /rkunix. You should protect it:

```
chmod 444 rkunix
```

On our system we created a file named LD, which contains

```
% ld -x 1.o m34.o queue.o c.o . ./lib1 . ./lib2
  chmod 444 a.out
  size a.out /unix
  mv a.out /unix
```

These commands load the object files shown to create a new system named "unix." Note that new files have been created: m34.o for the 11/34 and queue.o, a new queuing algorithm.

(4) Make special files for all devices. For a teleprinter:

/etc/mknod /dev/tty2 c 0 2

See mknod (VI). You may do ls – 1 on /dev to see what devices are there. Make sure that the device type and major code number match up with definitions contained in c.c and l.s. Note that a new device will need to have a file system constructed. See mkfs(VIII).

(5) Fix /etc/ttys, an ASCII file in which a one in the first column indicates whether a device is enabled or disabled. The second column indicates tty line. Editing the file and placing a 1 in the 1st column will activate the line. The /etc/ttys file will look like

```
000
110
120
0a0
1b0
```

(6) Type sync before bringing the system down.

(7) Now when booting you can type "rkunix" (the name of the system you created) to bring in the system. Of course, other names can be used to identify the system. From halt on 11/34:

(a) load SR: 173030 single user, 111111 multiple user
(b) control halt
(c) control init
(d) control boot

$ appears on console

(e) type RK <return>

@ appears

(f) type rkunix

\# appears if single user
% appears if multiple user

Appendix B
Modification of the MINIUNIX* System for Use on PDP-11/03s

B.1 METHODS FOR CHANGING ORIGINAL CODE

The MINIUNIX system is a small-sized UNIX* system provided by the Bell System that will run on PDP-11s without memory management. Supporting up to four users, the MINIUNIX system was originally configured to run on 11/40s. Basically, the MINIUNIX system features essentially the same attributes as the UNIX system but in a reduced form. To run the MINIUNIX system on a PDP-11/03, several changes are necessary. Most modifications needed are related to the distinct nature of the processor status register (PS) on the 11/03 where the PS does not exist as an addressable register. For example:

PDP/11 code	LSI/11 code
bis $50,PS	mtps $50
mov (sp)+,PS	mtps (sp)+

Changes of this type must be made throughout the file "mch.s" in order to allow it to run on the LSI-11.

There are also about 10 references to the "PS" in the code in other parts of the operating system. These references may be handled by calling an assembly-language function to implement the operation. For example,

* UNIX and MINIUNIX are trademarks of Bell Laboratories.

one may define the functions "getps" and "putps" to allow access to the PS.

Example PDP/11 code		Replacement LSI/11 code
ps=PS->integer;	->	getps(&ps);
PS->integer=ps;	->	putps(&ps);

The term "PS->integer" refers to the actual memory location of the PS. "getps" and "putps" are defined by the following "as" assembler program:

```
.globl      -getps, -putps
-getps:     mov 2(sp), r1
            mfps (r1)
            rts pc
-putps:     mov 2(sp), r1
            mtps (r1)
            rts pc
```

Finally, since the LSI-11/03 line time clock has no status register, references to the clock were removed from the main code and from the code concerned with the clock.

The tty handler was modified extensively, but only to improve performance. Other than the "PS" modifications mentioned, no part of the handler had to be modified to make the MINIUNIX system run on the 11/03.

The MINUNIX system is delivered on magnetic tape. It will be necessary to make the above modifications and insert new handlers prior to running on an 11/03. The simplest method is to compile a new system on another working UNIX or MINIUNIX system installation. Another alternative is to bring up the MINIUNIX system from tape on an 11/40, modify the code, and transfer the code to the 11/03. The most direct method of transferring code is to install a common peripheral device on both systems, e.g., floppy disk or hard disk. Referenced in Appendix C are two handlers written for two common disks that are used with PDP-11/03s—the AED 6200LP floppy disk and the DEC RL01. The user with other devices can follow the concepts outlined in these handlers for other common drives. Device handlers are added to lib2 when the system is generated as outlined in documentation supplied with the MINIUNIX operating system. Devices can be added easily, e.g., for the floppy disk and RL01:

```
% cc -c -O fd.c r101.c
% ld -x -r fd.o r101.o
% mv a.out ../lib2
```

Appendix C
Description of a Selection of Programs Available for Use on Laboratory PDP-11 Systems Using the UNIX* Operating System and 'C'

This appendix lists a number of useful program listings that are available from the author in a supplement to this book. Device handlers for devices not included with the original UNIX and MINIUNIX* system distributions are available as are programs for accomplishing the intercomputer communications described in Chapter 5. The listings for the device handlers will be provided only to holders of a UNIX operating system license. A tape of the programs is also available from the author for a nominal handling and materials charge. Programs that contain code specifically related to the internal code of the UNIX or MINIUNIX system cannot be distributed without a copy of the user's UNIX/MINIUNIX operating system license. Distribution media will be on 800 bpi 9-track magnetic tape in 'tp' format.

C.1 HANDLERS FOR UNIX AND MINI/MicroUNIX SYSTEMS

(a) UNIX. The following handlers were written at Vanderbilt University and operate with Version VI of the UNIX system

* UNIX and MINIUNIX are trademarks of Bell Laboratories.

(1) dh.c. DH-11. An asynchronous multiplexer that connects the PDP-11 with 16 serial communication lines.

(2) dz.c. DZ-11. An asynchronous multiplexer that connects the PDP-11 with 8 serial communication lines.

(3) si.c. System Industry 9400-62 controller for 80 mb disk drive manufactured by CDC (9762).

(4) la180.c. LA180. 180 char/sec printer.

(5) hp.c. RK06. 14 MB disk drive. The RK07 has twice as much storage capacity as the RK06. The handlers should be almost identical.

(6) fd.c Floppy disk for the Unibus. AED 6200-LP, Advanced Electronic Design.

(7) ip.c Handler for DR-11 for Interprocessor transfer.

(8) rk.c A revised RK05 handler based on the handler in the original UNIX system distribution.

(b) MINIUNIX system. The following handlers have been successfully used with a MicroUNIX system adaptation of the MINIUNIX system.

(1) dx.c. RX01. Single density disk drives manufactured by DEC.

(2) fd.c. AED-6200LP Unibus-compatible floppy disk drives used on 11/03 using a univerter. fd.c is the same program listed above for use with UNIX systems.

(3) r101.c RL01 DEC drive.

(c) TU-58 handler for either UNIX or MINI/MicroUNIX system. A complete user's guide for use of the TU-58 on the 11/03 or on UNIX systems is available.

C.2 PROGRAMS TO FACILITATE INTERPROCESSOR COMMUNICATIONS

(a) Serial transmission.

(1) vtty.mac. This program was written in Macro to provide communication between an 11/03 running RT-11 and a UNIX-based system. It operates in virtual terminal mode when first run, allowing the user to communicate directly with the host computer. Programs can be downloaded to the 11/03 into a file named 03.DAT (see Chapter 5).

(2) to03vt.c This is a 'C' program used for transferring a file from a PDP-11 UNIX system to an 11/03.

(3) out-sav.c A program used to convert the UNIX system header to

an RT-11 header so that programs prepared under the UNIX system can run under RT-11. Converts a file named a.out to a.sav.

(4) 11snd.c and 11rcv.c Two 'C' programs for transferring files on serial lines between two UNIX-based systems.

(b) Parallel communications. The following programs were developed for using the DR-11 and DRV-11 parallel DI/O cards.

(1) A handler for the DR-11 or DRV-11 is available as noted above.

(2) trans.c A program to transfer data or files over parallel lines using the ip.c handler.

(3) ROM programs. A set of programs are available to allow burning of a ROM containing a transfer handler for the 11/03. The programs for burning a ROM using a PROLOG ROM Programmer (series 92: 2708 chips) are available as well as the literature on the transfer programs that live in the ROM. An 11/03 with the ROMs in a MRV-11 will allow direct loading of programs from a UNIX-based PDP-11 system into memory.

C.3 EDITORS

(a) teco. The teco editor as used with RT-11, RSX and TOPS-10 has been converted to run under UNIX systems.

(b) edit. The "edit" editor distributed with RT-11 has been converted to run with UNIX systems. An RT-11 or other DEC operating system license (which provides teco or edit) will be necessary to provide UNIX-teco or UNIX-edit to interested parties.

C.4 PLOTTING

A complete plotting package for the Versatec 1200A printer/plotter has been developed. User routines for the 'C' language programmer as well as DMA handlers are available.

Index